Above the Line

Above the Line

Living and Leading with Heart

Stephen Klemich and Mara Klemich, PhD

HARPER
BUSINESS

An Imprint of HarperCollins*Publishers*

HarperCollins books may be purchased for educational, business, or sales promotional use. For information, please email the Special Markets Department at SPsales@harpercollins.com.

FIRST EDITION

Line drawings by Kirsten Agius

Heart illustration by str33t cat/Shutterstock, Inc.

Charts by Kristin Agius

Library of Congress Cataloging-in-Publication Data.

Names: Klemich, Stephen, author. | Klemich, Mara, author.
Title: Above the line: living and leading with heart / Stephen and Mara Klemich.
Description: First edition. | New York, NY: HarperBusiness, [2019] | Includes bibliographical references and index. | Identifiers: LCCN 2019027315 (print) | LCCN 2019027316(ebook) | ISBN 9780062886835 (hardcover) | ISBN 9780062886842(ebook)
Subjects: LCSH: Leadership—Psychological aspects. | Character. | Self-realization. Conduct of life.
Classification: LCC BF637.L4 K546 2019(print) | LCC BF637.L4(ebook) | DDC 158/.4—dc23

20 21 22 23 24 LSC 10 9 8 7 6 5 4 3 2 1

We dedicate this book to all of you
who step into courage to discover the best
version of yourself, in your very own way,
to make a difference in the world around you.

CONTENTS

FOREWORD

A few years ago, I was invited to speak at a client's world head-quarters in Texas to share my Heart-Led Leader message with the senior management team. After my talk, the CEO introduced me to two people who have been instrumental in developing the culture of this multinational organization over the last three de-cades. Little did I know that these two individuals would not only become my lifelong friends and mentors, but they would also change my life, both personally and professionally.

Stephen and Mara Klemich have been in the business of helping people find their best versions of themselves for more than thirty years. In that time, they have observed over and over that people can get trapped by their own fears, which in turn stops them from realizing the wonderful potential inside them. Stephen and Mara wondered, what if they could share some wisdom that could help people really understand what *drives* their patterns of behavior? In eighteen years of research and development, they went deep, examining and discovering *why* we do the things we do. What happens to us when we are separated from love—from the security of our authentic self? What happens when we live life out of coping strategies that cause us to *self*-protect and *self*-promote? What happens to our relationships, promotion opportunities, leadership and parenting style, and corporate culture when we operate out of self-protection or self-promotion?

Their goal was to design an instrument to bring some structure and language to what people intuitively understood. At the heart of our behavior there is a motivation that is based on four universal principles of life: courageous humility, growth-driven love, ego-driven pride, and self-limiting fear. And after years of statistical processes they developed the Heartstyles Indicator. A game changer.

Stephen and Mara have seen how these four universal principles of life are applicable across cultures and belief systems. They've seen it happen again and again, from CEO to the front line: when an organization introduces this wisdom into their culture, the positive impact on engagement and results follows. Astonishingly, sales go up, retention rises, and customer satisfaction increases. This work is applicable for both personal and professional life; it is relevant in sports, education, family, and business.

Stephen and Mara's book will make a difference for you right from the very first pages. There's good reason to expect that will happen. By the time you read this, more than 100,000 people will have participated in the Heartstyles program. It is now available in twenty-five languages, facilitated by certified practitioners all over the globe from Russia to China and Romania to India, through different cultures, economies, and belief systems—because these principles are truly universal. It's universal to *people* because it's about being *human*.

For the last two years, I've invited Stephen and Mara to share their Heartstyles message and Indicator at our annual Heart-Led Leader Retreats. I've seen firsthand the impact Heartstyles has on individual leaders and organizational cultures of all sizes. I've also seen firsthand the impact Heartstyles has had on my life, my marriage, my family, and my relationships at work. Stephen and Mara are not only master teachers, they are two of the most authentic and genuine human beings I've ever met.

I believe Stephen and Mara's book, research, and Heartstyles Indicator will begin a heart revolution throughout organizations across the globe. I believe we all can slip below the line and live in fear and ego. But Heart-Led Leaders push themselves to live and lead above the line. To live and lead with love and humility. We're all on this journey together, so let's get on this heart revolution together. And let's become our best version of ourselves so we can make a positive impact on the lives of others.

—Tommy Spaulding, *New York Times* bestselling author of
The Heart-Led Leader

Introduction

A Note on the Human Heart

On the day of the opening ceremony of the 2000 Sydney Olympics, thousands upon thousands of people lined the streets ten deep to watch the Olympic torch come toward them and go past in just a few seconds. The groundswell of cheering, clapping, and emotion intensified as the torch came past, held high in the hand of the runner. There was a sense that race, religion, and country didn't matter—people's hearts were united and caught up in the Olympic spirit. Spectators were polite and friendly, happily chatting with complete strangers, enjoying the uplifting atmosphere. As the flame was passed on, it was like the spirit of unity was passed on. Each runner ended their section, lit the flame of the next runner's torch with their own, hugged, high-fived, and did a little dance of joy. Runners completing their run were surrounded by excited family and friends as they moved away from the crowd and went off to celebrate. For the first time in Olympic history, runners were able to keep their torch (which they purchased to raise funds for charity). It was fascinating to see people wanting to physically touch a torch, as if something of that inspirational spirit could be passed to them by touching it.

Stephen was extremely fortunate to be selected as one of those runners carrying the flame on the day of the Opening Ceremony, as the flame made its way to the Olympic stadium where Catherine Freeman would light the cauldron. Stephen was privileged to be part of a bond of humanity not many get to experience in life. On that day, the hearts of humanity were harmonious. But as the crowds dispersed back to their day-to-day life, how easy it was to lose that sense of uplifted spirit and unity. In the hustle and bustle of finding a way back to cars and getting out of the inevitable traffic jams, how quickly we as people can shift from our patient polite self to our frustrated impatient self. One minute we're cheering and clapping for the unity of the Olympics and humanity, the next minute we're yelling with impatience at the traffic jam we are all in, in the hurry of needing to be somewhere. In both situations our heart, the core of us, feels the experience—from the joy to the frustration.

Deep inside your heart there runs a line. It is an exceptionally fine line—so fine that most times you may not even be aware of it. And yet the way we live our lives hangs in the balance along this line. How quickly our heart, what we refer to as our character, can go from one side of that line to the other side.

The wonderful thing is that the spirit of our best self is within us, within our DNA, within our hearts, our character. We as people can be part of creating positive and harmonious environments in our families, our workplace, our sports teams. We, too, can create an "Olympic spirit"—it just takes shifting our heart attitude, thus our thinking, and thus our behavior to make the world a better place.

When we talk about your heart, we are referring to the deepest expression of your character, the place where you feel love, fear, anxiety, excitement, anger, depression—all that rich and often conflicting swirl of human experience. The heart is the keeper of our emotions, a source of wisdom, and the deepest storage place of our values and spirituality. *Our outward behavior is a manifestation of the inward reality of the heart.*

When we begin recognizing the heart as the birthplace and seat of our desires and most compelling longings, we realize that it produces the elements, the raw materials that guide our life. Your rational mind will deconstruct the heart's messages to weigh up options, analyze the facts, and produce a set of guiding strategies that seem socially and logically acceptable. But the bread-crumb trail of your actions and behavior will always lead back to the desires and convictions of your heart.

The heart is where your authentic self, the golden treasures of your best self, can be found, discovered, and released. As you journey through *Above the Line*, we will be your mountain guides, leading you to reach great "summits" in your life, finding the gold within your heart for yourself and then those around you. We will equip you to live and lead *above the line* and move into your potential, becoming the person you know you can be.

Naturally, the line of life has two sides. One side is what we call the "above the line" side of your heart, filled with the kind of courageous humility and growth-driven love that bring out the best in us. On the other side—the "below the line" side—lie self-limiting fear and ego-driven pride that, despite all of our good intentions, bring out the worst in us and others. In a split second, every one of us can move from one side of the line to the other, all in the same sentence, behavior, or action.

Intellectually and intuitively you already know the line. When we ask participants on our programs to get a piece of paper, draw a horizontal line in the middle of the page, and then populate behaviors they would see above the line and below the line, we consistently get a page filled with two kinds of traits. Above the line traits include kindness, compassion, generosity, acceptance; below the line has aggression, hostility, avoidance, blame. Regardless of gender, age, occupation, or nation, participants instinctively "get" what we mean.

While you may be surprised by how naturally this exercise flows, you may also notice that just because you subconsciously hold this knowledge doesn't mean that you consistently live this way. The difference between what you *want* to do and the actions you end up taking can leave you thinking:

Why do I keep doing this? Every. Single. Time?

Why does it so often feel like the best of me is at war with the worst of me?

What is really going on for me and the people around me?

These are the questions the two of us spend our lives answering with teams and individuals around the world. Through thirty years of studying, coaching, practicing, and researching, we have found one truth to be consistent. *The above the line and below the line behaviors are based on four universal principles of life: humility, love, pride, and fear.*

Sounds simple, right? Yet in this world that so often operates out of negative fear and negative pride, all of us find ourselves moving below the line into our less-than-effective selves in order to cope with day-to-day life.

The wonderful thing is we're all designed for greatness, created to live above the line, to become the best version of ourselves in the way we live and lead, the way we parent, conduct relationships, do business, relate as a family, play competitive sports, and, of course, make a difference to the people and the world around us.

What we're really talking about is *character*. Living above the line builds and strengthens our character to manage life well; living below the line traps us into coping strategies and defense strategies. While these may give us a feeling of being safe and secure, they don't actually *build* our character, instead keeping us trapped in a false sense of security.

Our hearts are packed with good intentions, desires, and dreams, which is why it can be so *dis*heartening when we're triggered into betraying those good intentions. Each harsh response, cutting remark, or less-than-honest moment can come with a wave of regret and a wish for a do-over. Wouldn't it be great to have a "control Z" function in life like we do on our computers? Every single one of us has at some point said or thought, *I really could have handled* *that better.* The question that comes next, even if it's subconscious, is: *So why didn't I? And why did I do the same thing last week, or last month, or last year?*

We have heard these questions so many times over the years, from people aged seventeen to seventy. The setting is often different—a work meeting, around the dining table, or when one of "those" e-mails or text messages lands—but the frustration and confusion are the same. People feel baffled by their own behavior or the behavior of family, friends, and colleagues. They tell us:

- "I wish I could tell my boss no, but I said yes—*again*—and I'm burning out."

- "A few people at work have told me they think my sarcasm is harsh, but I'm just joking. I don't understand the problem."

- "My sister hasn't spoken to me in a year, ever since we argued about her partner's competitiveness at the bowling night. I miss her."

- "I want to spend a fun evening with my kids, but most nights, I end up getting on them about something."

- "I can't get through to a couple of people on my team. If they don't improve, I'll be forced to let them go, but they just don't seem to get it."

Welcome to the line, where so often the best of yourself is pitted against the worst of yourself. Our society, workplaces, and pressures of life are pulling our character below the line, while our authentic self, the greater character within, is pulling us above the line. This is the positive or negative tension of life. Life can be suspended in *positive* tension when you can see it, define it, and learn how to master it. We can become so used to living in *negative* tension that our heart can become blind to it and our less-than-effective character and behavior has a negative impact on others and our potential. We can succumb to pressure and stress or triggers that cause us to react in a way we really don't want to.

Every book on successful habits will tell you that we always have a choice in how we react or respond. We believe, though, that until you recognize *why* you do what you do, how the heart and the brain work together to shape your behavior, it will be hard to choose differently. It is our hope that this book will guide you through understanding the why, and help point you toward discovering the answers to your own unique questions.

We Are All an AND

Perfectionism is a self-destructive and addictive belief system that fuels this primary thought: If I look perfect, and do everything perfectly, I can avoid or minimize the painful feelings of shame, judgment, and blame.

—BRENÉ BROWN, *THE GIFTS OF IMPERFECTION: LET GO OF WHO YOU THINK YOU'RE SUPPOSED TO BE AND EMBRACE WHO YOU ARE*

Warm and hostile, kind and prickly, hospitable and frosty . . . we humans are remarkably contradictory creatures. Not one of us is all good or bad, 100 percent love or 100 percent ego. We are all of these things. We can live below the line and above the line in the same day, hour, or even sentence!

This is what we call the power of the AND. We are all an AND because we are all in the process of *becoming*. Becoming the person we often find ourselves wishing we could be. Well, guess what? It is possible—but yes, it's a *journey*. That word has been used a lot over the years. Yet it's true—life *is* a journey! And in that journey, our best self and our less-than-best self coexist. That's why we all know people (and let's face it, ourselves, too) who can be incredibly annoying, tiresome, and scary; yet in another context they can also be caring, kind, and thoughtful. We are all an AND. If you see yourself and others that way, you can have a lot more compassion for yourself and others, extend more grace, and be more patient and tolerant as we all walk this path of life.

Not so long ago, we were told this great story about the power of the AND. Two couples who had experienced our teaching went for a ski vacation together. For the first three days, one of the guys continually talked about himself with a bigger, better story: he wasn't listening to any other conversations, not asking anything about the other couple, interrupting and swinging every topic back to himself. It was three days of "me, me, me" and no "we." Eventually over dinner, the other guy in his frustration said, "Okay mate, we know it's all about you. All you've done for three days is talk about yourself, *your* business, *your* life." You can imagine how that sarcastic tone of voice threw a chill on the evening: so much so that everyone went to bed early with an egg beater of emotions churning in their gut. Both wives ended the night saying, "That dinner conversation went well—*not!*" The next morning at breakfast, Mr. Talkative came to the table and apologized. He admitted he had been offended by his friend's comment the previous night, but realized it was true. He'd been through some tough work issues recently and was not feeling so confident about himself. As a result, he was talking himself up to his friends, who he knew always encouraged him. He admitted he'd been so self-absorbed that it *had* become all about him. Then his friend stepped in and admitted, "Mate, I was totally below the line last night in the way I came at you with sarcasm, instead of being more authentic and asking what was happening for you. I should've realized that you weren't being your usual caring self, so there must be something going on for you. I know your heart and you were out of character." They shared a man-hug and laughed about how quickly great friendships can come under tension and how thin the line can be at times.

Courage Is Not the Absence of Fear

There once was a young man under extreme tension who was full of conviction but, driven by fear and the rejection of his community, believed that aggression was the only way to bring about change. He was imprisoned for his involvement in violent political protests, and for nearly three decades he languished in prison, enduring the most brutal of conditions. You can imagine that when he was finally released, this man (no longer young) would turn his thoughts to revenge against the regime that had so tortured him. And yet that was not what he did. Instead, he walked out of those prison doors with the strength of character, the humility and love, to lead: using his power not to take revenge, but to lead a nation into peace. This man's name, of course, was Nelson Mandela.

What happened to Nelson Mandela within the prison walls? What shifted within him during those long years? How could he go from leading with aggression and anger to leading with courage, humility, and love?

Nelson Mandela walked a thin line. It was the line between racial hatred and loving all people; the line between a negative aggressive approach and a humble assertive approach; the line between seeking power for his ego and seeking it for his nation.

There is no passion to be found playing small—in settling for a life that is less than the one you are capable of living.

I learned that courage was not the absence of fear, but the triumph over it. The brave man is not he who does not feel afraid, but he who conquers that fear.

—NELSON MANDELA

There is a Nelson Mandela in all of us, burning with the desire to make a difference.

We can be locked in the fortress of self-limiting fear or the prison of ego-driven pride, or we can choose to walk out of it, to shift to courageous humility and growth-driven love, to use our character for good.

You might not set out to be a Mandela and change a nation, but by living an above the line life there's no doubt that you will change *your* world around you. Sometimes that's just about getting along with your mother better, managing your team under business pressure, not blasting your horn at that driver who cut you off, or not picking a fight with the man in the express line at the grocery store who clearly does *not* have twelve items or less.

The working title for this book was originally *The Four Universal Principles of Life*. These principles—humility, love, pride, and fear—ground what you will read in ancient wisdom, but also resonate deeply with modern neuroscientific discovery. Our eighteen years of research and development is summarized as behavioral philosophy: understanding the logic that underpins behavior, and what it means to live and lead with heart. We then set out to research and discover the behaviors that manifest from the four principles and formed them into a model that became the Heartstyles Indicator. It's now available in twenty-five languages, and we have seen this philosophy taught to diverse audiences across the globe, ranging from CEOs in corporate boardrooms to frontline workers in far-flung regions to village chiefs in remote African communities (yes, really!). These principles are both universal and timeless.

Part 1 of *Above the Line* will take you through the *why* and *what*. It explains *why* people behave the way they do and *what* above the line and below the line behaviors look like, so you can identify them and be equipped with a "compass for life." In part 2 we share many *how-to* skills with you. Chapter 4 also includes a free MyPack Self-Score Heartstyles Indicator (or HSi for short) for you to complete. You will be joining more than 100,000 respondents who have stepped through this process, experiencing the effectiveness of above the line behavior in their personal and professional lives. It's important to know that this Indicator is very different from the personality tests you may have done before. This is a *life* indicator, not a *type* indicator—a character development tool, not a personality profile. It describes what your heart attitudes, your thinking, your behavior, and your life look like right now, and creates a benchmark of how you would like them to look in the future, as opposed to creating a static label for "who you are" based on personality traits.

On top of that, there are guided activities, as well as a QR code in the appendix in the back of the book to access workbook pages, view video learning, and download our app to connect with us on your mobile device. Anywhere you see an asterisk * in the text, that means you can access additional material through that QR code in the appendix.

The message of living and leading with heart has been a long time evolving. At the beginning of 1994, we had spent several years facilitating leadership, team, and personal development programs for organizations. It was the year of an important revelation for us: the people we saw who were able to grow and improve always began that journey with a change of *heart attitude,* while others didn't develop even though they agreed with the principles being taught. This is the kind of heart attitude change you often see when people go through traumatic situations like being diagnosed with high blood pressure and then giving up smoking, or a massive financial loss in an organization and then cutting back on wasteful expenses, or a divorce that forces both partners to take a good look in the mirror and change their behavior. It was then we both realized: character development and the ability to become more effective in life starts in the heart.

Early in the process of designing the Heartstyles Indicator, we found ourselves sitting at a table with a group of senior executives in Switzerland on a leadership program. We were having a philosophical discussion about what makes us human and what drives our behavior. They were a group of highly educated, successful executives with very diverse backgrounds, nationalities, and spiritual beliefs: Muslim, Hindu, Buddhist, atheist, agnostic, New Age, and Christian. Stephen described our research, model, and behavioral philosophy: that each of us is driven by four universal/spiritual principles—humility, love, pride, and fear. Each in their own way said, "Mara and Klem [Stephen's nickname], you've got something very special here. We all agree with this philosophy. These are universal principles. We've used other instruments that show *how* people behave, but this is finally giving us the *why*." They agreed these four universal principles transcended any barriers of belief system, culture, language, worldview, or religion. From that day on, we knew we had a model that would cross all cultures and beliefs, and we set about researching and developing it.

We all agreed that living life above the line brings fulfillment. We also all agreed that it is, of course, entirely possible to win and succeed by living below the line. Push your way forward, compete at any cost (these are the aggressive strategies), or avoid any conflict or disagreement, as it could end relationships or careers (these are the passive strategies). What "lies" below the line are, indeed, lies: deceits you have been told or have come to believe are the only way to get ahead and survive in life. On the surface, below the line aggression can work, especially in the brutal corporate, civic, and political world we live in. Remember the schoolyard, and how bullies ruled that world? Aggression provides its own sense of power, status, security, and a way to survive life. Likewise, passive fear-based strategies can seem to stave off trouble for a long time. But both kinds of below the line strategies are proven to be ineffective in the long term. They destroy relationships and even our culture. They come with a high price tag: divorce, estrangement, bitterness, brokenness—even war. So it leaves us with a choice to live and lead below the line or above the line.

As you journey through this book, it will provide you with insight for transformation and help you understand how the world works in terms of people and their behavior. We will show you how to navigate the line and the changes and challenges of life that come your way, personally and professionally. We will equip you to understand what's going on for you and for others, and how to handle the pressures and stress this high-energy, fast-paced world has for us. We will offer a framework and a language for you to be able to identify behavior so you can manage it more effectively, more of the time, with more people. We will do that by sharing stories, examples, and how-tos, so you, too, can experience transformation and live with more freedom, belief, and confidence—just like the thousands of other people we have seen transformed over the years of our work.

We will show you the difference between above the line and below the line behaviors, and give you insight into the data from more than 100,000 respondents across the world proving the effectiveness of above the line behavior in personal and professional life. You will walk away knowing the ageless, timeless wisdom of the four universal principles that drive behavior from the heart and be equipped with practical how-to tools that will enable you to live and lead above the line.

Here is our promise. By exploring what has happened to our hearts, how it shapes our thinking, and what our behavior actually looks like day to day, we develop an incredible strength and power to choose more effective behaviors, especially when we're triggered. But it doesn't stop there. As we grow our own character, we can better support and strengthen the people around us through empathy, understanding, support, and modeling excellence and achievement. That is how each of us can change the world for the better: from the heart. It's ageless and timeless wisdom. We call it the Heart Revolution, and it is our ultimate goal.

So fasten your seat belt—this is an invitation to join us on an epic adventure as we serve and guide you on a life-changing, life-developing, and life-giving experience.

Staying Safe on Your Mountain Climb of Life

TTP!—Trust the Process

Getting lost on our "mountain climb of life" can happen to us all. Usually when you're literally going somewhere and you can't find your way, you can stop and look for the "you are here" icon on your map or device, or find a sign that tells you where you are. Then you look for where you want to be, and you plan how to get there. In the same way, this book will give you a map, a "compass for life," as one of our clients called it many years ago. As you put one foot in front of the other, navigate around the crevasses of life, and learn when to keep going and when to stop and take some time to consider where you're at, we can assure you all things work together for good—even if you can't see it at the time. We always say *Trust the process—TTP!* Just because we're feeling lost today doesn't mean we will be that way tomorrow. The process of growth has ups and downs, and it is a process—so TTP!

As part of the work we do in our leadership and personal development programs, we've guided people up mountains in more than ten countries and kept them safe. As mountain guides, be it rock climbing, abseiling, or alpine mountaineering, we have an advantage of knowing where the summit is. In life, though, the summit keeps moving and even growing. But just as we have guided people up mountains (and down again), we will also guide you through the material we have used successfully with thousands and thousands of people. Like any good guidebook, this book isn't based on what we will do in the future, but what we already know works: the ideas and methods that we have tried and tested over thirty years of facilitating people in their transformation.

Climbing mountains is all about patience and endurance, placing one foot in front of the other. As it is in life and our transformation journey. To climb well, you need to have the right equipment, together with a map, a compass, a plan, the skills, and what we call the smarts. Above all, it takes heart to climb well. The heart of courage, respect for the environment, a willingness to bury the ego (ego-driven mountaineer = death), to keep believing in what you can do even when your body is tired, to have resilience and discipline. Once again, it's just like life. So, you will capture in this book the heart attitudes that will bring you to your life summits, and you will also learn the smarts and the skills to assist you on the path. We call it the "Heart + Smart" equation.

You Can't Be Brave if You're Not Scared

More than twenty years ago, on a Himalayan peak opposite Mount Everest, after a year of planning, weeks getting to base camp, and days on the mountain, we were three hundred feet from the summit when Stephen turned the team around. He was confident in his decision, as the mountain conditions were becoming very dangerous. The weather was changing rapidly, and he took into account the experience level of the team and himself. On that day, after consulting with his Nepalese Sherpa Tschering, wisdom prevailed, and Stephen said, "Today is not going to be summit day."

We didn't summit that peak on that expedition: the storms came in and stayed for days. All the team had personally invested a lot of time, energy, focus, and commitment to training and readying themselves for that expedition (Mara often said that if she had to walk up and down yet *another* set of stairs with a big pack on for training, she'd break something—or someone). We were all, in our individual ways, really wanting to reach the summit. It wasn't an easy decision to say no to the summit—but it was the wise one. Silencing the ego will open your heart and mind to different perspectives and truth. If Stephen had listened to his ego and the team's attachment to the summit, we would have blinded ourselves to the imminent dangers and the ending would have been dire.

That experience of silencing the ego and turning around was as brave as making it to the summit. If we hadn't turned back, we could all still be up there, our corpses frozen in history. Since then we have had countless adventures all over the world and we're always grateful for summits when they happen. Two decades later, with much more experience and with better weather conditions, we *could* summit if we went back. As you pursue the "best you," know that some days are not summit days, just as some months or years are not summit years, because you have to grow in heart and smarts to reach your individual great heights of life. Yes, at times it is scary, and that is why so many do not reach their potential. When you get to the mountain, it is never exactly how you thought

it would be. It never matches what Google says or the online reviews. Sure, it's *something* like what your research tells you, but
what brings the most joy is your own bravery in getting up there
on that mountain and navigating it. Remember TTP!

You're Not Alone

Sometimes in life, solitude is required to find yourself, to meditate,
pray, reenergize, center yourself. Sometimes, if you're climbing,
you might even do a solo climb. But most of the time you're climbing mountains with a buddy, at least one other climber. We're not
meant to do it alone. We suggest you read this book with a very
close friend if possible. Discuss with each other what you are discovering, uncovering, unlocking. Connect with us on social media
to get with others on the journey.

Another time in the mountains Stephen thought he was going to
die from hypothermia. Dramatic, but true. "I was doing a charity
adventure event called the 15 Peaks Challenge in Wales, which, as
the name indicates, requires you to climb fifteen peaks in twenty-
four hours. Having set out at four a.m. in the dark, after nineteen
hours of climbing, it was now eleven at night. Our team of four was
trapped in the eye of a most severe storm off the Northern Atlantic,
which ended up flooding most of England. Eventually we reached
the summit of one of the peaks, though we could hardly stand up,
as the wind and rain were so strong. It was like standing directly in
front of a fireman's hose on full blast. All four of us were soaked to
the bone; our core body temperature had dropped so severely we
were shaking uncontrollably with hypothermia. One of us managed to text the local Welsh Mountain Rescue team. It took them
more than twenty minutes to respond with compass bearings and
distances to travel: that felt like a lifetime in those conditions, and
we were thinking, *If we don't find a way out of this tonight, we will
die here.*

"For us to navigate out, we had to turn back and hike in the direction from which we had come. Psychologically, that was just this side of impossible. We had done all we could to reach this place where we finally found cell phone connection, and *now we were being told to go back the way we came. What??!!* But we had to trust the process. We had to trust the compass and the directions, the rescue team, and each other. At least if we could get moving, we had some momentum, we had some hope—sitting on that summit doing nothing but shivering, we were hope*less*. Very slowly we got moving, and after an hour and a half in pitch black darkness and rain slashing our bodies, we saw in the distance about a dozen flashlights flickering. It was the mountain rescue team, who had come out in those horrible conditions to guide us to safety."

Whether you're wanting to continue your growth journey, or when you're in a desperate situation in life, we hope that this book will be one of the compasses that will help you navigate; we hope you can reach out to friends and maybe even professionals to guide you. There was no shame in Stephen needing professional help from the mountain rescue team, even though he is very experienced in the mountains. There is no shame in talking to someone like Mara for professional psychological help, no matter what your profession or place in life.

We have watched people from all walks of life have their hearts healed from past wounds, their character strengthened, and mindsets changed from *can't do* to *can do*. They have learned what shaped their life, thinking, and thus behavior, and they have identified the strongholds that have held them back. They have learned how to rise above the ineffective behaviors to a life that is whole, purposeful, empowered, and making a difference to the world around them.

Adventure creates great energy, and great energy creates more great energy. So climb the mountain with us: be courageous, share in the adventure, and experience the energy that comes from life-transforming climbs. "Rope up" with us and we will guide you on the adventure of the Heartstyles mountain, taking delight in reaching greater heights in your own life and the lives of those around you.

We hope you find a way to connect with other people on the

same journey, the same mountain, to encourage you, to climb with you as life grows and changes, and you find purpose and meaning in making the world a better place.

As Tschering Sherpa says, the journey of life can be "a little bit up, a little bit down, and a little bit flat." We invite you to enjoy the climb of a lifetime . . . and TTP!

PART I

Why We Do the Things We Do

CHAPTER 1

The Four Universal Principles That Shape Your Life

As you begin this journey, is there a longing within you? "Yes!" you might think. "I want to hit my quarterly sales goal," or "I want to lose ten pounds," or "I want to stop arguing with my spouse about how to load the dishwasher." Or maybe you pause, reach in a little deeper, and think, "I want to feel fulfilled," or "I want my relationship with my partner to be better," or "I want to raise happy kids," or "I want to find my soul mate." These thoughts, big and small, about our daily life and desires reflect a greater longing within all of us.

Deeper within, we are all seeking more fulfillment, purpose, confidence, contentment—ultimately, *more love*. In our pursuit of this, we live our lives in ways that we believe will bring us "more." We try to understand why other people think and behave the way they do, and why *we* behave the way we do. We wonder why other people's less-than-positive or helpful behaviors are so obvious to us, but not them. And we question why it is that we sometimes resort to our own less-than-positive behaviors, or how we can avoid taking the low road next time something presses all the wrong buttons.

In all of this, what we are really seeking is wisdom about what motivates people (ourselves included), so we can be the best versions of ourselves and achieve our personal best. What if we could find that wisdom—easily, without taking a vow of silence for a year or meditating on a mountaintop? (Which, depending on the season and the mountain, could actually be quite pleasant.) What if it was actually right there, in plain sight?

Well, it is: that wisdom really is right in front of us. It is unlocked when you recognize the line, what lies above it and below it, and when you understand the four universal principles of life. Together these insights can help you achieve the "more" you've been longing for, the "more" you are designed for. They can help you be your best self, most of the time. And they can help you successfully navigate all the glorious, sticky situations you encounter and build a happy, successful, fulfilling life.

Sarah was hoping for some of that wisdom as she sat in front of her boss's desk, staring at the pattern in the carpet. *Don't just sit there—say something!* her self-talk was shouting. But she couldn't do it.

When she had left home early to catch the subway, she was feeling good about the coming workday. The feeling was crushed as soon as she walked into the office and her boss, the VP of product development, caught her in the hallway. "Can you come into my office?" His tone and those six words dragged Sarah's happy attitude down into the doom zone.

Her heart rate was high as her boss questioned her about her current project, the biggest of her career. "The end of the quarter is a week away, Sarah! Why is this project dragging? It's not like you have a lot of other things on your plate. Our revenue goal for the year depends on us releasing in the third quarter. Is there even a chance of that?"

Fifteen minutes later, when she walked out of the office, Sarah wasn't sure how she had responded. All she seemed to hear was her blood rushing in her head. She had a vague memory of promises to get things back on track and a mumbled explanation of some of the unexpected hurdles the team had faced. She felt like breaking something.

As she slumped into her chair, her phone rang. It was Simon, her husband. "Honey," he said as soon as she picked up, "I'm sorry, but I need you to come home right away. I locked myself out of the house while I was putting the garbage out, and I'm going to be late for work."

On another day, Sarah might have been a little frustrated but laughed. On this day, though, it was the least funny thing imaginable. She felt a swell of anger, took a breath, and let Simon have it. She resorted to nasty language. She accused him of doing this type of thing *all* the time. "Why do I *always* have to clean up after you?!" Then she told Simon how his problem was going to impact the deadline she could see looming on the horizon—how *his* problem was going to ruin *her* month.

Simon's silence stretched on and on. "You know," he finally said, "you sound just like your father."

Sarah almost threw her phone down. Instead she jabbed the "end call" button—twice.

Alice, her friend and counterpart in market analysis, arrived at her office just as she was stalking out. Studying Sarah's grim expression, she asked, "Are you okay?"

"Yeah, great. Why hasn't your team given us the market report yet? You told me we'd have it last week. Are you *trying* to sink this project?" As the words came out, Alice literally took a step back, her eyes growing wider.

Sarah saw her response, but it was too late to change course. She felt the anger that had driven her to speak to Alice that way. A flood of regret and embarrassment filled the void—she didn't know how to recover. "I'm sorry," she blurted, "but we need the report now or I'm going to catch it from Bill."

Alice's face was blank as she explained the report would be ready by the end of the day, and then she left. Sarah thought about going after her to offer a real apology and tell her about the meeting with Bill, but her feet didn't move. *I have to get home*, she thought.

On the subway, Sarah couldn't stop thinking about what Simon had said. *You sound just like your father.* She knew, in her heart, he was right. She had opened her mouth and her dad, with all his blame and bitterness, had poured out. But why? Why had she treated Simon and Alice, two people she cared about, the way she had? And why hadn't she spoken up to Bill about the engineering team being pulled in too many directions—their biggest hurdle, and one Bill could help solve?

Every day, most of us have moments like this. *Why did I say that?* we think. Or, *If I had just . . .* We're sharp with our kids or partners over small things. We criticize our teammate in front of the others. We agree to a deadline we know isn't realistic. Or maybe the issues are bigger. Maybe we've lied about something important. Maybe we've taken an unethical shortcut, cheated or lied. *These behaviors don't make us bad; they simply make us human.* They are the coping strategies we use to survive in life—and they've been with us a *long* time. Often they are grounded in good intentions that are turned upside down by our less-than-effective coping strategies—all because we have connected to fear. *What?!* You may well ask. *It's not because I'm afraid!* But we all can be, and much of the time we don't even realize what we're doing, let alone why.

Here's the good news: every day, we also make our *best intentions* a reality. Within ten minutes of wishing we could pull words back into our mouths or make a different choice, we can be supportive, focused, honest, patient, and committed. How quickly the heart can shift from selfish to selfless, from judging to compassionate, motivated to depressed, constructive to destructive, full of doubt to confident. We can be effective one minute and ineffective the next.

We are all an AND. *Life* is an AND. Ineffective, below the line behavior coexists with effective, above the line behavior, and we are all able to switch from one to the other and back again in the blink of an eye.

That AND is the essence of the line that exists within our heart, and the four universal principles of life* that drive our behavior. They are:

- Courageous HUMILITY—focusing on personal growth
- Growth-driven LOVE—focusing on growing others
- Ego-driven PRIDE—focusing on self-promoting
- Self-limiting FEAR—focusing on self-protecting

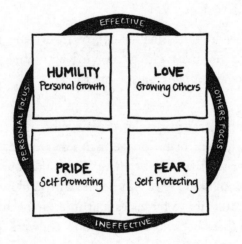

In its entirety, the model represents the circle of life, and on any given day, moment, or situation our heart can go around this circle—from fear to pride to humility to love. Around and around!

Our life is an outward manifestation of the reality of the heart.

- When the heart is operating out of self-limiting fear, it displays self-protecting thoughts and behaviors and it limits us with behaviors like passivity, dependence, and the need to please.

- When our heart is operating out of ego-driven pride, it displays self-promoting thoughts and behaviors. Through behaviors like aggressiveness, superiority, perfectionism, and winning at all costs, it blocks us from really connecting with others. Certainly there is a positive pride that is an expression of love—the delight we take in people, our personal achievements, and things we are pleased or thrilled by—but negative pride places importance on our ego: proving, performing, power, or control.

- When our heart is operating out of courageous humility, it displays thoughts and behaviors of personal growth through behaviors like authenticity, diligence, vulnerability, and achievement.

- And when it's operating out of growth-driven love, it displays thoughts and behaviors of growing with others through respect, loyalty, honor, and compassion.

The four universal principles explain what drives the incredible range of human behavior we manifest around that circle of life. We can watch the same behaviors play out around the planet, and repeat throughout history, in people with completely different backgrounds and life experiences and in entirely different situations. We see these behaviors daily in how we work, how we play sports and games, how we parent, and how we build (or can damage) relationships. They turn up in the movies we make and the books we write and the art we create. They shape our families, communities, teams, and organizations.

Some aspects of humility are:

Courageous   Strength of Character

Life Giving — HUMILITY — No Ego

Safe Energy   Selfless

Some expressions of love are:

You may attribute the way you behave to emotional patterns such as anger, envy, guilt, compassion, kindness, or generosity. Yet in our eighteen-year journey of research to understand why we do the things we do, spending more time with statistical modeling than any reasonable person might choose, what we found was that all these different motivators can be traced back to just four original sources. Courageous humility. Growth-driven love. Ego-driven pride. Self-limiting fear.

It is these four universal principles that explain why the best aspects of your character are sometimes pushed to the background—by self-limiting fear and ego-driven pride—and why your great inner strengths of character are released—by courageous humility and growth-driven love—so that your best self wins out.

Our character is not our *personality*. Character is the inner strength to live out of values such as humility and love especially when under adverse situations, to keep calm command in stressful situations, and to remain positive when everything else around you is negative. Character is also the courage to think and behave above the line when everyone and everything else is below the line.

The important thing to know is that every one of us is at times operating out of fear or pride, reverting to coping strategies when

we perceive any form of threat to our security emotionally or physically. That may be when we lose our job, get rejected by someone,
offend someone, or experience a negative impact on a relationship.
Anything we do when driven by fear or pride won't be fully effective, but it's a normal way to react. It might deliver some short-term
results, it might even feel good in the moment, but it won't actually
help us grow our character. It can be hard to see this happening in
our lives, though, because fear drives pride, and pride drives denial.
And when you're in denial, you're blind to what has brought you
there.

While we universally value humility and love, and all the inner
and outer benefits they bring, not one of us is *always* driven by
them. There is no such perfect person. That does not mean that we
have to live our lives below the line, though. We don't have to be
trapped in the fortress of fear and the prison of pride. We can shift
above the line.

The month before Sarah's less-than-best self morning, she had
been part of a workshop we were facilitating for leaders in her organization, where we explored the four principles. As she sat on
the subway, she recognized the below the line principles that were
operating in her life—the pride that had led her to blast Alice and
Simon, but most specifically, the fear she had felt in Bill's office
that was driving her behavior. She wanted to move away from that
feeling of fear, and so she focused on shifting her heart, her thinking, and her behavior. By the time she got home, love had risen

above the fear. She immediately apologized to Simon and told him she loved him and that he hadn't deserved to be treated that way. And she told him about her morning meeting. Simon apologized for adding stress to an already difficult day. She left the house feeling more peaceful and connected, and ready to get back to work, apologize to Alice, and achieve what she needed to for the day.

We have all been where Sarah was that morning, but we don't have to be ruled by our fears and ego-driven pride the next time. The first step in understanding why we do the things we do is to understand the power of these four principles in our lives, and how they form the root of our behavior—and that of others. When we do, we become more self-aware, more powerful through our positive choices, and our best self more of the time as a partner, parent, professional, leader, and friend.

The Heart of the Matter Is Always a Matter of the Heart

The heart turns up in conversation all the time. Follow your heart. Her heart is in the right place. He had a change of heart. Her heart wasn't in it. He was all heart. They had a heart-to-heart. My heart goes out to him. He broke her heart. It was a half-hearted effort. I mean it from the bottom of my heart.

We talk about the heart every day. It's part of our natural vocabulary. For thousands of years, we have spoken of it as more than just a pump. But have you ever really thought about what it means? When we say somebody spoke from the heart, it means they spoke with meaning, insight, and sincerity. Or that the deeper reality of each of us is reflected when we're following our heart.

This is why we say that *the heart of the matter is always a matter of the heart*, and that the four universal principles live within the heart, arranged in matching pairs above and below a central line. They are the deeper reality of our character, our thinking, and our behavior. They are a source of insight and clarity about how we each work, love, and live. Understanding what is happening in somebody else's heart is how we feel compassion or empathy.

Love is our greatest need. Rejection is our greatest fear. We spend our lives seeking love and avoiding rejection. As John Lennon once said, "There are two basic motivating forces: fear and love. When we are afraid, we pull back from life. When we are in love, we open to all that life has to offer with passion, excitement, and acceptance." We feel these two drivers in our own behavior and recognize them in the behavior of others. If we can grasp this wisdom, it will guide us to effective life relationships and successful leadership.

As you begin to climb the above the line mountain, we'd like to spend a few minutes with you at base camp, guiding you through a mindfulness exercise to experience the difference in what happens for you when your heart is in a place of love or in a place of fear. In our busy lives we often don't take the time to stop and contemplate where we are at. So find yourself one minute—only one minute—to go through this simple yet profound exercise. By *consciously* focusing on love, and recognizing how different it feels from fear, you can feel yourself shifting to a calmer place. It helps us know we *can* shift our emotional and physical state. This is empowering!

Below the Line: Fear and Pride

As facilitators working with people globally, both of us live out of a suitcase for a large chunk of our time. With offices and clients around the world, we have spent more than thirty weeks out of the year moving from one hotel room to another, and go for long periods of time without setting foot in our headquarters (or our home). We arrive, we work with our teams on projects, we come to agreements on timelines and deliverables, and we're off again.

EXERCISE: ONE MINUTE AT BASE CAMP

You can do this exercise in most day-to-day situations in life. Wherever you're reading this book, you can do this exercise right now.

1. Close your eyes and for twenty seconds, connect to the love in your heart. Recall a time when you felt loved or were loving. Who was there? What was happening? Really see it and feel it. Where do you feel that love? What's it like? What happens for you and to you as you feel it? (Know that this is purpose, connection to others, the strength of your relationships, your belief in yourself. It is the real you.)

2. Next, turn your heart toward fear. We all have it within us; none of us are immune. Recall a time when you felt fear. Who was there? What was happening? For these twenty seconds, step into that memory, really see and feel it. Where do you feel that fear? What's it like? What happens for you and to you as you feel it? Feel the insecurity, the doubt, how it makes you need to protect yourself or prove yourself. (This is the real you, too, but it isn't the best you. It's the you when you are in the grip of fear and pride.)

3. Finally, for the last twenty seconds, surrender yourself to love again. Feel the difference. What does that love show you about yourself?

That is the journey of this book: acknowledging that we all feel fear and pride and that they influence us daily, but love is what we want most, and humility and love together are the path to unlocking the best within you—your true strength of character, your best version of yourself.

A couple of years ago, we had to deliver an important presentation for the C-suite team of a client organization. One of our team, Ted, jumped at the chance to create it. "Leave it with me," he said, "I can handle it." Stephen gave him a thorough briefing, checked that Ted understood and felt capable of proceeding, and left for his next trip. Over the next three weeks, Stephen checked in with Ted to find out if he needed anything. Ted assured him that the presentation was coming along and to "leave it with him." Stephen arrived back in town, tired, jet-lagged, with a huge number of appointments in his calendar for the week, one of them with Ted. When Ted started to share what he had done, it was obvious that he didn't have the experience he had claimed in creating the type of high-level presentation we needed. Very little information had been gathered, and what had been completed was not high quality. Ted was floundering.

Stephen, in the grip of fear and pride, was stressed and angry. "Ted, this is just not good enough! I've given you three weeks to get this done." He wasn't shouting, but his tone was abrupt, his frustration was unmistakable, and his negative energy could be felt a mile away. The office was open-plan, and about five other team members were in the room with them. All five of them wanted to hide under their desks.

Ted didn't look up.

"Ted, please look at me."

Ted raised his gaze, but his stare was flat, his expression blank. "I've been asking people to give me what I need to get this done, and they haven't. Nobody was helping me."

"We might not be able to deliver to our client because of this. We have to fix it—now." Stephen could feel that things were spiraling, and so he told Ted to send him everything he had, then walked back to his desk, not feeling any better and knowing full well that he had broken a cardinal rule of leadership: praise in public, coach in private. We also still needed to get the presentation done, but how effective was Ted going to be now?

Stephen and Ted were both *below the line*, behaving in ineffective ways that were driven by fear or pride. What was happening in their hearts in that situation?

The presentation was for an important client who would have boosted our company and our brand. When that felt at risk, Stephen slipped right into fear. Fear acted like a vise grip on his character, putting him in self-protection mode—protection from the heart's ultimate fear, which is rejection—and preventing him from being his best. For all of us, this is true: when we are locked in the fortress of fear, we hide. We hide who we are behind a mask, and by presenting the person we think others want to see, we rob them of the chance to meet the real, wonderful person within. We hide from the truth, from difficulty, from honest conversations, even from possibilities. We develop tunnel vision and become blind to the opportunities for positive change all around us. We resort to ineffective coping strategies for dealing with our negative emotions and thoughts rather than proactive choices about how to best achieve our goals.

Those ineffective efforts to deal with fear often give rise to pride, in the negative, ego-driven sense. At the root of ego-driven pride is fear. *You're rejecting me! I'm not good enough, what if I fail? I'll show you.* In the prison of pride, we learn to manipulate our way around situations in life to feed the ego. Ted's fear of being caught out on the fact that he hadn't been truthful about his capabilities caused him to fall back into pride and to try to defend himself by blaming others. His responses triggered Stephen's own pride-based coping strategies: he became focused on regaining control and proving his ability to make things happen and achieve results. Lack of confidence is a form of fear, which we most often mask with aggression, itself a form of ego-driven pride.

The difficulty for many stuck in pride is that it is the issue behind the issue and can be difficult to recognize within ourselves—or others. Pride can drive feelings of self-doubt when we're wrong or when we make a mistake, which can break down our self-worth. It can cause us to adopt a spirit of blame, which is certainly where Ted was, or superiority, which limits or damages our relationships, where Stephen was. When we're operating out of pride, we block ourselves and others from reaching our potential, even when it looks like we're successfully achieving our goals. You know that pride is in the house when things get stressful and exhausting.

Have you been in this situation, either in Ted's shoes or Stephen's, as a leader or team member, as a parent or partner or family member? We're pretty sure the honest answer for all of us is *yes*. This is life. We all fall below the line sometimes, and we have all been sucked into the downward spiral. Think about the last argument you had with your spouse or a family member or a friend that seemed to spiral down until you had both said things that were too harsh, maybe that you didn't even mean.

The question we always ask is "How's that working for you?" Instead of being sucked into the spiral, the best prevention strategy is to strengthen our hearts and our character, which we'll write more about in the next chapter and in chapter 3.

It may not surprise you to learn that people who report that they have high levels of stress, low levels of happiness, poor effectiveness at work, or low-quality relationships also score themselves quite high in pride and fear-based behaviors—and so do the people around them who complete the 360-degree version of the survey. When you complete a Heartstyles Indicator, you can get a Self response on how you see yourself, and you can have the option to ask other people to rate you on the same questions, so you get how others see you, too. It's about how others experience you—and sometimes our intentions are not communicated to others in the way we expect. People's Benchmark (their aspirational survey results) tend to show how much they want to be operating out of humility and love more often.

Above the Line: Humility and Love

"What do you mean you don't know!? What are we paying you for? You're *supposed* to *know*!" Paul's blood was boiling, his voice full of frustration. This was the third meeting he and his sales team had had about this situation in the last ten days, and each time the atmosphere in Paul's office increased in anxiety. Why wouldn't this senior leader sitting in front of him speak up and step up? It was his job to deliver results—and he was just sitting there, looking at the floor, with no good explanation as to why his group's numbers were still sliding.

"Well, go and find out, then!" Paul dismissed the leader and his team member in a voice that revealed his Bronx accent and his working-class roots the more frustrated he became. As the two of them gathered their things and sped out of his office, Paul sat down heavily in his chair and pondered what was going on. Why was he surrounded by incompetents? Why did it take yelling at people for anything to get done right? Fueled by this latest skirmish, he set off for the next meeting that, he was sure, would hold yet another absence of ideas for how to change the direction of the company.

Paul was a highly intelligent man who had grown up with the belief that if he wanted to do well in life, he would have to fight every step of the way. No silver spoon for him. If he wanted to get ahead, he had to be better than others. And so that's what he did. He got top grades, he won at sports, and he beat others out for scholarships. He got his first job in the competitive financial industry in Manhattan. He vied for promotions. He drove his teams hard to prove he could deliver results.

When he stepped into his current position as CEO of a global cosmetic company, things weren't in good shape. The company looked glamorous on the outside, but inside there was a desperation to turn the numbers around, and people were operating out of fear and pride, trying to survive. Paul was *determined* to turn it around. He demanded more effort. He pushed leaders. He didn't see anything wrong with using aggressive tactics in the workplace—being ruthless got results.

The problem was, he wasn't getting results. People had stopped coming up with decent ideas. More and more, he heard whispers in the hallways and blame in meetings. Fear of failure had set in and ego pride had become the norm—drive, control, and don't dare make a mistake. If you do make a mistake, lie about it or blame someone or something else. Some of the company's best talent was leaving. They were embroiled in a lawsuit that they couldn't seem to resolve after years of fighting. And their numbers continued to fall. But Paul just kept showing up ready to fight it out, and his below the line behavior was pulling everybody else below the line with him.

One day, his chief people officer appeared in his office. Paul didn't want to talk about people issues when he had major operational issues to deal with, but it was clear the CPO wasn't going to leave until she had her say. "Our culture is going south. It's becoming toxic. The infighting is going to kill us if we don't fix it."

Paul sighed. "Fine. What do *you* want to do about it?"

"It has to start with *us*, with the leadership team."

The day we came in to work with Paul and his senior leaders, we could feel the fear and tension in the room. The team looked at us with suspicion, like the whole day was going to be one long test. Paul was barely engaged. But by the end, a few candid comments and discussions had signaled a minor shift.

After our first workshop, when we saw small breakthroughs but mostly an intense desire from everybody to turn the business around, we started coaching Paul. What we quickly learned about him was that the person he was at work was entirely different from the person he was at home. He was a wonderful husband and father, loving and playful with his kids. He was a beloved volunteer coach for his eleven-year-old son's soccer team and a patient teacher. He did everything he could to be there for his family, to be supportive and to show how much he cared, despite the demands of his job. We coached Paul to change how he saw himself in the office and to bring more of his "father's heart" to work—in how he led and how he treated his team.

Why were there two different Pauls? Along his path of success, he had made the subconscious decision that "out there"—at work—you had to be tough, and at home is where you felt and showed love. He had compartmentalized his "real self" into his personal life, and believed his "work self" had to be somebody quite different. Like most of us, he was reluctant to admit any doubts in front of his colleagues. "Our culture values strength and power, and showing fear is considered weakness," says Leon Hoffman, codirector of the Pacella Research Center at the New York Psychoanalytic Society and Institute in Manhattan. "But you are actually stronger if you can acknowledge fear." We suggested to Paul that if he wanted to change the culture, which could dramatically change the results of the business, he had the opportunity to rethink his leadership style. He could bring more of his true self to work.

Paul wanted to turn the business around, not just for results but also because a lot of people's jobs were at stake. He wanted the company to enjoy the success they used to have in the market, and he didn't want to live like this, but he knew no other way. When he began to understand that the fear and ego pride–based culture in the company was ineffective and was getting in the way of results, he had to own up to being a large part of the cause. He had to shift away from his own fear of "not making it" and the aggressive pride-based behaviors that he was using if he wanted to develop a different culture that would propel the company into growth.

Rather than think of his employees as people he had to control, he chose to think of them as people he could coach and develop, just as he did with his soccer team. Paul realized he knew the boys on his son's soccer team more than he knew his own leadership team. His relationships with his colleagues were only transactional. He came to understand that he needed to think about his team the same way he thought about his soccer team: starting with compassion and a belief that everyone was doing their best and it was his role to support and coach them. So he started to connect individually with each of them, being vulnerable about his own fears for the business and how that was impacting his leadership and personal style. He apologized to each one for his ineffective behaviors, admitting he had been operating out of a lot of fear, concerned his job was on the line if he didn't get results. The change in Paul's attitude was so genuine, he created a safe place for his team to share their own fears and concerns.

He started small. He stopped interrupting or finishing people's sentences. He managed his voice tone, didn't lean forward, point, or chop one hand into another when he spoke. He admitted when he didn't have the answers and accepted ideas that were better than his. He found time to sincerely recognize and praise great work. Most importantly, Paul became intentional about being honest with himself and continually picked himself back up when he fell into fear- or pride-based thinking and behavior. He asked his team to support him in his growth by calling him out on any below the line behavior—and he accepted the feedback. Though, he said, sometimes with gritted teeth!

It took a few months for his team and the wider company to believe that this new, caring person was real. Slowly, Paul and his leadership team built true and trusting relationships, and that spilled down into their departments. As they became encouragers and developers, supporting people rather than instilling fear in them, the company's fortunes started to change. Employees became engaged in the brand and passionately supportive of their colleagues and their leaders. This led to more innovation, more boldness in suggesting new ideas, more productive meetings, and a generally more effective company culture of "we're all in it together" and "if it's meant to be it's up to me." As the culture has improved, the company's financial position has improved right along with it.

And each day, Paul is continuing to become a much happier and more fulfilled person.

That is what we can achieve for ourselves and others when we are *above the line*, behaving in effective ways that are driven by humility and love. No matter where we are—at home, at work, playing soccer, on vacation—our hearts are longing to love and be loved. When we can surrender to that and understand the power of it, it can change our lives and free us to change other people's lives. Strange though it may seem to talk about love in the context of our workplace, there is what you might call "companionate love," that quality that comes through when colleagues care about each other's work and even non-work issues. This kind of love not only feels good, it boosts employee morale, customer satisfaction, and teamwork.

This is ageless and timeless wisdom. From love, we unlock the values of respect, honor, and compassion. We connect with others more deeply. We live more fulfilled lives. But to do so, we can discover and identify our own inner value, and that actually emerges from humility.

You don't hear people talking about humility much these days. It may bring to mind a sort of false modesty. For a definition of authentic humility, though, it's hard to go past this from Steven Sandage, director of research at Boston University's Albert and Jessie Danielsen Institute, who is studying this trait. Humility, he says, is "realistic self-awareness of one's strengths and limitations, the capacity to regulate emotions of shame and pride, and a concern for others."

Here's what humility looks like. It's the ability to admit you're being driven by fear and pride, which then unlocks genuine courage. The way Brené Brown describes vulnerability in her TED Talk "The Power of Vulnerability" (one of the most viewed TED Talks of all time) and authenticity in her book *The Gifts of Imperfection*—both of those words are another way of describing humility. Through this kind of humility, we can be courageous enough to be honest about our need for personal growth as well as our own strength and inner worth. Humility dethrones pride in our lives and allows us to face difficult situations with calm command rather than the need to prove and control.

How Humility and Love Release Our Potential

With humility and love, we live *from* significance—our inner worth and value—rather than *for* significance—seeking our worth from our external environment by proving, performing, and perfecting. In our society, humility and love have been associated with people who are wishy-washy, weak, naive, lacking in grit and determination, too nice, or too emotional in our world of competitiveness, aggression, and ambition. The truth is that people who operate with humility make others feel safe, respected, and cared for, and that builds trust and connection *with accountability*. This is the foundation of greatness, of accomplishment, of achievement.

As you might expect from what we have described so far, according to the data from our Heartstyles Indicator, those who scored high in love and humility behaviors report high levels of happiness and effectiveness at work, high-quality relationships, and low levels of stress. Now that's an outcome worth pursuing!

We are all doing the best we can with the wisdom and character strength we have in the moment. What's so exciting and liberating is that we can shift our heart, our thinking, and our behavior so that we are more often above the line. We can all do this by connecting with the humility and love in our hearts, especially when we have already been below the line.

Getting to the Far Side of Denial

Denial is a natural outcome of pride, and to get above the line, we can choose to develop the self-awareness to recognize it and then make different choices. That's what Stephen chose to do in his interaction with Ted. After Stephen left the office, he spent time thinking about how to get back above the line. He asked what had made him behave the way he had. He needed to address his behavior with the team, in an authentic way, and that wouldn't be possible if he couldn't acknowledge that it was driven by what was happening in his heart.

The next morning, he pulled the team together. "I'm sorry for my behavior yesterday," he told them. "It was below the line, and not my best self, as you know." He went on to explain his perspective. "I was tired, just coming off the back of an intense facilitation trip of giving out to everyone else, and I lost my composure. I criticized Ted in public, and that wasn't right. I should have had more compassion for him and taken him aside to address his performance in a calm way in private. It was my bad, and I'm so sorry."

His apology was sincere because he was actively working from a place of humility, and the team knew it. This was the Stephen they all knew, most of the time!

That was a start, but he still needed to coach Ted. And he was still frustrated and deeply concerned about the presentation.

It's easy to think that we'll end up below the line whenever we're angry, but that's not necessarily true. Most days in our lives, we are going to experience negative emotions when things don't go well: anger, disappointment, frustration, hurt. In those moments, we can still connect with our best intentions. We can go where love is, in our hearts, to help elevate our behavior above the line. We can focus on what we truly want to achieve, beyond ourselves and our self-interest, not what we're trying to prove. Stephen spent a lot of time thinking about what he wanted for Ted—to help him be his best, with compassion and respect for who he was as a person, and to continue to coach him to give him the opportunity to further develop in his role. He reflected on what he knew of Ted as a person, and what triggers might have caused Ted to behave the way he had. And he considered how working to help Ted grow could help us spread our message. With these feelings and thoughts driving Stephen, he could now have a constructive coaching conversation.

In that conversation, Stephen acknowledged his below the line behavior and apologized. Then he asked questions in order to understand what was happening for Ted, in his heart. Ted finally admitted that he was afraid because he had overestimated his ability and had been trying to bluff his way through. Stephen shared his perceptions in a truthful but compassionate way and then turned the discussion to what would be right moving forward, rather than focusing on who was right in the past. Together, they decided on next steps. At the end of that meeting, Ted stood up, hugged Stephen, and said, "Thank you for your understanding, and for being willing to help me develop. I'm really grateful."

We've found that your energy introduces you before you speak. You've probably experienced someone coming into a room with positive or negative energy, and you've picked up on it before they say or do anything. Stephen started off with negative energy from fear and pride, and thus created a negative atmosphere. When he shifted to humility and love by apologizing, he changed the atmosphere to a positive and safe place. We all have a choice of changing our own attitude and behavior, and that then can impact our circumstances.

Not long ago, we got an e-mail from Sarah. She told us about an interaction with her boss. Bill was questioning her about progress, and Sarah felt herself shifting into fear and avoidance. Because she could now recognize and acknowledge it, she had the courage to address it. Instead of staying silent, she chose to be authentic and was straightforward about the challenges her team was encountering and talked about the struggles of sharing resources with other teams and other projects. Bill was surprised—it was somewhat new information for him—but supportive. He called a meeting with the heads of other teams to work on priorities. Things became a little less stressful and more productive.

Can we control every facet of our emotions and our lives? No. On any given day, from one hour to the next or from morning to afternoon, our heart can shift. As our environment, our circumstances, our interactions, and even bigger aspects of our lives change, different principles are awakened within our heart, influencing our thoughts and shaping our behavior. We might be driven above the line one moment and below the line the next. But that does not mean we don't have a choice.

When we can recognize the four principles operating in our lives, we are empowered to make better choices that lead to stronger relationships, greater confidence, clearer vision, and more purposeful achievement. Imagine what it might look like in your family, in your company, in your community if more people made those types of choices every day. The possibilities are infinite.

CHAPTER 2

Triggers, Templates, and Truths

Having explored the principles that drive behavior, we will now unpack what shapes the *thinking* that causes our behavior. You will be equipped to better identify and understand the causal factors of *why* we behave the way we do.

Eva, her two kids, and her husband were sitting on the living room floor around the coffee table, laughing at a story Eva had just shared from her recent travels. It was a Friday night, music was playing in the background, and the scene was set for quality family time. But the happy tone of the evening was about to shift.

Eva's ten-year-old son, Sam, was getting upset. "But Mom, I won't be able to buy a hotel!" he said, waving at the Monopoly board between them. He had landed just three stops short of GO.

"You're on my property. You have to pay," she replied.

"Can I pay you on my next turn?"

The question annoyed her. "It's a game, Sam—the whole point is to win."

Jessie, the twelve-year-old, rolled her eyes, which irritated Eva even more.

"It's just one turn. I promise I'll pay."

"No! You have to pay me *now*."

When Eva heard the hard edge in her voice, her tone of frustration, even anger, a part of her cringed inside. She wasn't surprised when Sam's eyes filled with tears or when he got up and ran to his room. *Why?* she thought, as her heart sank. *Why did I just do that?*

She had been away on a business trip for ten days, hopping from Vietnam to Cambodia to Thailand, then a long overnight flight home. It had been exhausting, but successful. She had been tough—really tough—in negotiations with suppliers and felt she had "won some points" with her boss along the way. She just *knew* her quarterly numbers would beat those of her colleagues on the sales team. But in the taxi on the way home, all she could think about was hugging her kids and having a relaxed, fun evening with the family.

"Let's play Monopoly," she'd suggested after dinner—it seemed like a good idea at the time. How had her good intentions turned so quickly into such a disaster?

Weeks later, Eva was still mulling that night over, still feeling the regret in her heart. At a retreat with her peers we were leading, exploring this question of why we do the things we do, she broke down. As the story came out, she paused for a moment and said, "I just *couldn't let him win.* Deep down, I knew in the moment it wasn't right, that it didn't matter. All I wanted was to be closer to him after being away. But I *couldn't stop myself.*"

We all have opinions about letting kids win, but ask yourself, was Eva really trying to teach Sam a lesson about resilience in that moment? Was that her heartfelt intention? No. She was still subconsciously in work mind-set, trying to win, and her ten-year-old was in her way. The more important question is, why was she behaving so competitively, to the point of frustration and arguments, with her son when what she really wanted was to spend a loving evening together? Why *couldn't* she see the minor disaster unfolding in front of her and stop herself before it was too late?

Why do we do the things we do?

In chapter 1, we explored the significance of the line that divides effective from ineffective behavior, and we encountered the four universal principles of life—humility, love, pride, and fear—that drive our behavior. But how do they do that? It's fairly clear that Eva's heart was operating out of pride in that moment with Sam, but how did what was happening in her heart translate into her behavior? And just as important, why was pride the dominating principle for her in that moment?

We hear about the importance of self-awareness and emotional intelligence—in leadership, in parenting, in marriages, in all our relationships and endeavors. The people we have watched unlock their greatest potential are able to live self-aware of heart (emotions) and head (mind-set), because our behavior emerges from both. If we live unaware of either the heart—the core of our character—or the head—our thinking "control tower"—we miss the full picture and struggle to grow. Our capacity for positive transformation is infinite, but only if we can address those things that keep dragging us below the line.

When you understand why you do the things you do, you can make better choices. You can more easily avoid those moments that send pangs of shame and guilt to keep you awake at 2 a.m. You can achieve more of your goals, build stronger relationships, and lead a happier, more fulfilling, more purposeful life.

That happy place is exactly where Eva was longing to be, and the only way to begin the journey was to ask, *On that Friday night, what was happening for me? What were my inner thoughts? What was happening in my heart?* When she was given the tools to explore and answer those questions, she could recognize a pattern of behavior in her life and begin to shift it.

We can all do the same—starting now.

Understanding Your Behavior Patterns

"Next time, I'm *not* going to do that."

How often have you said this to yourself, but then "the next time" comes around and, sure enough, you find yourself doing the same thing. Or maybe you see the pattern in others. You watch a friend end every romantic relationship when it starts to get too "serious," or watch your boss snap at you and your colleagues more and more as the end of every quarter approaches.

We *all* repeat behaviors. Sure, we tend to focus on those we would like to *stop* repeating, but the wonderful news is that we also repeat our *best* behaviors. Maybe you rush to the aid of any friend in need or help your coworkers when they're struggling with tough projects. Repeating more of our effective behaviors, those moments when we know in our heart that we were our best selves, and fewer of our ineffective or negative ones is the key to feeling more confident, secure, connected, and fulfilled. But for that to happen, we can learn the wisdom to recognize the patterns playing out in our daily lives and understand where they come from.

That was Eva's struggle. Her heart was telling her something needed to shift, and that fateful night of family Monopoly made her realize that she didn't want there to be a "next time"—but she wasn't sure what to do about it. Eva hadn't yet connected the dots in her pattern of behavior and didn't know that it was being generated by a simple formula:*

$$\text{Situation} + \text{Thinking} = \text{Behavior}$$

$$S + T = B$$

But what does this formula really mean? Jack and Jill are walking through Central Park when a barking dog runs toward them. Jack stops in his tracks and turns to run away or pick up a stick to defend himself. Jill doesn't break stride and moves toward the dog to pat him and see if he's lost. Jack and Jill are in the same situation, but their behavior couldn't be more different. Jack is scared of dogs, having been bitten as a child. He thinks, *That dog is going to bite me!* Jill has had dogs as pets all her life, including one who was lost for a week when she was ten. She thinks, *That dog might need help.* Even though they were both in the same situation, each behaved differently based on their thinking about dogs.

Sounds like a simple, basic formula, right? Well, it *is* simple. And yet it's a profound key to living life as our best selves! We tend to attach a lot of the explanation (even blame) for our behavior to that S (situation), but this approach prevents us from seeing the same base patterns that show up again and again in what seem like very different situations.

What we also can discover is that there is *context* in life. Context is about how our past shapes us. We each have an individual context, and thus a unique worldview. S+T=B helps us identify the context of our previous and current thinking, and we can then amend our reactions.

Our heart knows things our mind can't explain. Our mind can live in denial, but our heart seeks truth. Jill and her friend Jack aren't stuck with their reactions. They have a choice. As the influential psychologist Rollo May once wrote, "Human freedom involves our capacity to pause between stimulus and response and, in that pause, to choose the one response toward which we wish to throw our weight." To take advantage of the opportunity to make a choice aligned with our true intentions, though, we need to explore our thinking, which is a combination of three Ts.

The Ts: Triggers, Templates, and Truths of the Brain

When *situations* in life present themselves, our triggers, templates, and truths kick in to shape our *thinking* and therefore our *behavior*.

Or to put it another way, our thinking emerges from our triggers, templates, and truths and creates our context.

$$\text{SITUATION} + \text{THINKING} = \text{BEHAVIOR}$$
$$\text{TRIGGERS}$$
$$\text{TEMPLATES}$$
$$\text{TRUTHS}$$

Triggers are cues in the environment (picked up by our five senses) that stimulate certain thought patterns (positive and negative). Those thought patterns are based on templates—each one a collection of stored memories, emotions, and sensations from past experiences that have been filed away in the brain and help us quickly process what's happening. Based on those experiences, we also develop truths, or deeply held beliefs about ourselves, others, and the way the world works. Jack and Jill were thinking very different conscious thoughts based on their unique triggers, templates, and truths. Jack: dogs bite = be scared! Jill: that dog may be lost, I need to help him = go and pat the dog. Each had a different context based on their past experiences and their truths they formed *from* their past experiences.

So are either of those thinking patterns *the* truth about dogs? No, but for Jack and Jill, based on their own life experiences, it is each one's truth—*my* truth. In reality, both may be true: some dogs *do* bite and some dogs *are* friendly. The key to living above the line is knowing and understanding what shaped our lives and what our triggers, templates, and truths are, so we can be enlightened to know how to respond and behave effectively.

Our truths based on negative experiences can become a trap that we live in—they can become stories we tell ourselves to cope with life. We can then operate out of "the story of my life," that is actually a story of "*my* truth," but not necessarily *the* truth. This can lead us to become trapped in a set of self-beliefs that block us from living our potential. Yet we all can be released from this trap to discover a *greater* truth of who we *can* be and live the life we *can* create.

Let's return to Eva to see how it plays out in more detail. With the Monopoly board between them (the situation), Sam asked his mom to let him delay paying the rent he owed. She thought, *The point is to win,* and so she demanded that he hand it over. In the midst of the situation, she had some conscious thought about what was happening, and then she behaved accordingly. But what prompted Eva to think as she did?

The game of Monopoly was, in some ways, very similar to the meetings Eva had been having with suppliers over the previous weeks, with her boss sitting next to her. It was important to her to "win" in those situations, and when Sam began to negotiate with her, it *triggered* the same emotions and thoughts. Suddenly, it seemed important for her to win in *this* situation. The irritation she felt was a clue that she had been triggered: sudden shifts in emotions are often signs that we have been triggered. In chapter 1, we saw Sarah being triggered by the meeting with her boss, Stephen being triggered by Ted's bluffing and blaming, and Paul being triggered by his team's inability to turn the company around as he thought they should.

But *why*? Why, for instance, did Eva feel so competitive with her son, Sam—and not just with Sam, but also with her colleagues, suppliers, and even her friends at times? Was it just work that triggered this competitive need to win in every situation, or was it a life pattern? If we look at Eva's life, what we learn is that she was the youngest of three kids, with two older brothers. "My whole childhood," she shared at the retreat, "I had to fight to win anything. Even at the dinner table to get food!" Her brothers were older, bigger, and stronger. If Eva wanted to win, she had to be faster, tougher. Her parents, who came from a lower middle-class background, had high expectations for all their kids, and to earn their attention, she believed, she had to do better, be better, and always win—and that shaped Eva's triggers, templates, and truths—at work, or playing Monopoly!

How Our Thinking Works—
Our Brain's Hard Drive

All of your experiences—good and bad—are stored in your brain, which uses them along with input from our current environment to shape your thinking and behavior. When those experiences are positive and associated with feelings of inner worth and respect, they tend to create templates and truths that lead to above the line behaviors, based in humility and love. When those experiences have been less than positive, associated with feelings of rejection or lack of worth, they tend to create templates and truths that lead to below the line behaviors. The templates from the negative experiences in our lives can be powerful, creating strong surges of emotion or physical reactions, especially because they are aligned with our very basic fight-or-flight instincts. As a result, we're driven to behave in ways that self-promote (pride) or self-protect (fear)—these are our *coping strategies.*

The specific template that Eva had developed was simple: *I have to win to be seen and to be approved of. And to win, I have to be tough.* Encoded within that template of promoting or proving her worth were the emotions and even physiological reactions associated with her early experiences: the anger she felt when her brothers laughed at her when she lost a game and the hurt when her parents praised her brother for his grades and said nothing about her achievements. (More than likely they did praise her, too, but somehow, it didn't feel that way to Eva—our brains are not always good interpreters of the truth!)

We call the limbic system, the part of the brain that regulates our emotions and reactions, the *hard drive of the brain,* because this is where our templates are stored. The situation is your finger that clicks on the mouse (the trigger), which selects the file where a relevant template is stored. When Eva was triggered, her limbic system clicked on and opened the file with her "be tough and win" template, then put it into action. In that split-second moment, her neocortex, which analyzes input from the limbic system and determines our thinking and behavior, was operating on the truth that winning is how you prove self-worth, and losing means a loss of worth in other people's eyes. Was it *the* truth? No. But it was Eva's truth in that moment, and in many others. The same template and truth play out every time Eva is triggered in this way, and this is how our thinking turns into patterns of behavior.

Wondering what special gift you can get for your children? Positive templates are the gifts that just keep on giving. When Elizabeth joined our team, we were amazed by this self-assured young woman. She was only twenty-three years old, with no previous experience working in the corporate world; before joining us, she had been an outdoor guide. Right from the start it was clear that Elizabeth had no fear of authority. She was never nervous or uncertain with her boss or anyone senior to her. She had a voice, she seemed to know just who she was and where she was going, and she appeared happy in her own skin. Never arrogant, Elizabeth was very teachable, always willing to learn from others. *How do you get to be like that?* we wondered. Once we learned a little about Elizabeth's childhood, it all made sense. She grew up in a family in which everyone had a voice. Each evening, everyone sitting at the dinner table took part in the conversation. From the positive templates of her childhood, Elizabeth had learned to be bold and unafraid, ready to speak and equally ready to listen. Her triggers, templates, and truths around people and authority were *I have a voice and I am significant.*

Using the Ts to Understand Others

Understanding how our patterns of behavior are formed can be incredibly valuable in our efforts to shape our own lives and achieve our big goals and greatest desires. However, understanding other people's patterns is important, too, if we want to have a positive influence on the world around us.

A friend of ours is a sixth-grade teacher, and she shared an inspiring story with us. Jacob was a new boy who had recently joined the school. He had been expelled from his previous schools and came in with a record of serious behavior issues. His low self-esteem was obvious, as was his effort to compensate by being defensive and "acting tough." He was often in fights with other boys in his class, verbal and physical; challenged the teachers often; and struggled to fit in or find friends. He was identified as a student who needed monitoring and assistance.

Our friend Stephanie had a lot of empathy for Jacob, because she could see that behind the tough-boy mask, he was in emotional pain. His parents were divorced and he lived with his mother and stepfather. His family life was dysfunctional, based on interactions between the school and his parents, and the boy seemed genuinely unhappy. She decided to work on building a rapport with Jacob. It took time, but Stephanie would have chats with him about what he'd done over the weekend, how things were going in his life. She showed an interest in him. If he challenged her, she would be firm but respectful and express her belief in him. He would often call himself stupid or dumb, and even shared that his mother said the same about him. Stephanie would counter that by telling him that he was intelligent and that he had a ton of potential. She would show Jacob that she trusted him by giving him responsibilities during special activities and events, which he accepted and performed well. She could see how proud he was in those moments, and she doubted he'd been given many opportunities to feel this way in the past.

But the distance between Jacob and many of the others in his grade, particularly a "popular" group of boys, continued. They began to deliberately provoke him, and he would respond aggressively. Often he would initiate out of revenge for something that had happened the day before. Tensions escalated and visits to the principal increased. The school tried to keep Jacob separate from the other boys, but that wasn't entirely possible. Discussions were held and consequences were dealt, but it was obvious to Stephanie that the boys had no real empathy for each other, their apologies weren't sincere, and the behaviors weren't changing.

She had heard about the S+T=B formula and triggers, templates, and truths, and after yet another major confrontation between the boys, she decided to apply that perspective to the situation. Stephanie held a meeting with all the boys, including Jacob's only friend at the school. As usual, each side blamed the other, and it became apparent that the latest incident had ignited because Jacob was in a "bad mood." When she probed as to the source, he admitted that he'd had a fight with his mother that morning. The other boys had reacted to his negative mood with aggression and it had spiraled out of control.

She explained S+T=B to the group and spent some time on triggers and templates and above the line and below the line behavior. She asked the boys to think of an emotional experience they had had, and whether they might have a template built upon it. One boy admitted that he was terrified of bees because he had been stung once. Other boys shared stories of accidents, holidays, and even movies that left impressions on them. She then asked them to think about times when they had been in "bad moods." What had caused them, and how had other people reacted? Did others know they were upset and why, or did they just judge and respond? Returning to the incident of the day, she asked whether they had done something similar to Jacob, and whether their behavior was below the line or above the line. Finally, there was a breakthrough and they began to show some genuine remorse.

The group continued exploring the choices we can make in how we respond to other people's templates and behaviors. Suddenly Jacob, who had been listening quietly rather than being defensive or argumentative, shouted out, "My stepdad bashed me up last night!" Everybody watched in shock as he broke down. Stephanie wrapped up the session quickly and escorted Jacob to the counselor's office, where he was offered appropriate support and procedures in line with child protection protocols were begun.

Before sending them back to class, Stephanie asked the other boys to keep what they had learned private, but to reflect on it and to consider how they might behave in his situation. Four of the boys made time over the following days to genuinely apologize to Jacob and made an effort to include him rather than reject him. Tensions eased.

When we understand why others might do what they do, we are given incredible insight, and a real chance to not only build our relationships in powerful ways, but also positively influence their lives. If sixth graders can do it, why shouldn't we as adults?!

Not all templates come from dark places. Ken, a senior leader in a corporate role, is clearly very high in compassionate. When he described his upbringing to us, Ken painted a picture of the only child of a single mother, raised in the projects with very little money to spare. His mom worked two or three jobs at a time to get by. Even as a young boy, Ken would look for ways to help his mother out. He figured if he skipped lunch at school, he could put his lunch money back in his mom's purse. He even went out of his way to make friends with kids who had the sort of parents his mom could connect with, so that she wouldn't feel so lonely. Not only was he deeply aware of his mother's material sacrifices, his empathy for her extended into an awareness of her emotional needs. What a beautiful template he formed in those early days of his life, and that has made him the courageous leader he is today. It allows him to keep people's best intentions top of mind, and start with the positives about them as a person, rather than reacting negatively when they are below the line. Ken told us about a situation that showed this: he and his team were in an intense problem-solving meeting when one team member, David, became very offended by some of the comments being made. David threw his pen down on the table, saying, "I've had enough of this!" and stormed out of the room, slamming the door. Some of the team were quick to openly criticize their colleague's behavior. Ken politely but firmly interrupted them, saying, "This behavior is out of character for David, and we need to compassionately find out what is happening for him, rather than criticizing him." Ken himself searched David out and supported him by saying, "You must be under a lot of stress at the moment—I know your behavior was out of character. I want you to know I've got your back." Creating that safe place allowed David to be honest about what was happening for him. To his credit, David was able to go back into the meeting and apologize to his colleagues. Ken's empathy and warmth are the foundation on which his achievement drive sits, so he is a leader who successfully balances getting things done with caring for others. All of this comes from his positive childhood templates.

The Overly Helpful Brain—
"It's Just a Movie!"

Have you ever wondered how you can sit in a movie theater and watch actors playing characters you know aren't real, doing things you know never happened, all on a 2-D or 3-D screen, and yet you can still laugh, cry, feel sick, or get angry? None of it is happening in real life, to you or to people you care about, but you still feel an incredible range of emotions and even physical sensations. Why is that?

The brain is "helping" you by pulling up your templates.* You might think, *It's so sad that character's grandmother died!* and *feel* that sadness, because based on just two senses—sight and sound—your brain is replaying the emotions and physical feelings you experienced months or years ago when somebody you cared about passed, or the empathy you have felt for someone else in that position. You might not be consciously thinking of that person, but your brain pulls up the old templates, says, "Close enough," and *overlays that experience onto the present moment.*

A few years ago, Mara and I were at a lovely beach resort. Mara had an awful food poisoning episode that week. Her template was: do NOT, ever again, eat prawn curry! Every time she thought about the resort afterward, she'd get a shivery feeling. That bad experience caused two templates: one—about the specific *dish*; and two—about that particular *place*. Even though the resort had been lovely (mostly the food was great!), that food poisoning and associated template colored Mara's memory of the *whole* experience.

Have you had a bad food experience and swore you'd NEVER eat that food again, no matter where you might encounter it; or an overly indulgent amount of a particular cocktail or liquor that now makes you shudder with revulsion anytime someone even mentions it? (If you smell it, that's the worst!) Or you went somewhere for a holiday and had a terrible experience, so your opinion of that particular hotel chain, and maybe also the entire city or country, is as low as it can go?

These are all examples of how our brain files away templates, then later on brings up similar templates, even though they may not exactly match the current situation. "Close enough" can be very unhelpful, because your current situation *is* different from what you experienced in the past. It's like the eager but ignorant guide who sends you along the wrong path down the mountain. Close enough might land you on the wrong side of the mountain!

Mara's shuddering and refusal to eat prawn curry would not be appropriate if she were at a friend's place for dinner, that friend is an excellent cook, and makes that same style of curry for dinner. If Mara didn't know about her template and how her brain works, she could ruin the moment by saying, "Oh no! I can't eat that! Just the smell makes me gag!" Not a good way to cement a friendship!

Every second of every day, your brain is "helping"—sometimes a bit too much!—by clicking on then pulling open the files on your hard drive, pulling out templates, and generating positive and negative emotions, physical sensations, and thoughts, all based on the input of our five senses. If your brain didn't store negative and positive templates as reference points, every day would be a new experience and you would be barely functional. You wouldn't be able to make decisions or use your wonderful imagination or be capable of compassion. Our templates make us capable, complex, caring, emotional human beings.

But those templates also spark emotions, thoughts, and sensations that may not be the right match for the current situation, just as Eva's did. We aren't sure who said it first, but we are fans of the saying "How we do anything is how we do everything." Our templates are the reason for our repeated patterns. Eva's "be competitive" template is how she plays games with her son, is how she negotiates with suppliers, is how she compares successes with her brothers over Christmas dinner. That template had seemed to serve her well over the years. She had graduated from high school at the top of her class, went to a prestigious university, and was recruited by a respected company. In her career, "playing to win" had helped her succeed—getting top sales numbers, winning the highest profile clients, negotiating tough terms. During her weeks of travel and negotiation, Eva's old template *seemed* to help her achieve her goals.

In the moment with Sam, when he became her rival rather than her son, it did not.

The word "seemed" is an important one here. While Eva was successful in many respects, the cost of that success, achieved via at least one pride-based template, was a lot of unnecessary stress, anxiety, and exhaustion. Eva could have achieved the same things with a heart operating out of humility and love and had far fewer moments like the one she had with Sam. Rather than *needing to prove* her ability to succeed, she could have been *confident* in her ability to succeed. To know herself and not need to keep proving, performing, perfecting—to love herself and find peace in who she is.

But it can be difficult to distinguish between below the line and above the line approaches to achieving our goals when we've been successful using those below the line approaches. It can be hard to see when a template is in play because our templates are also backed up by associated truths that are being broadcast by our neocortex, which is busily analyzing the data stored in the limbic system. If the limbic system is the storage center, think of the neocortex as the control tower, analyzing and making decisions on how to deal with all our emotions, memories, and our current situation. Its mission is to protect us from pain. Excellent! But its very dedication sometimes causes us problems. Every time the neocortex gets a message from the limbic system that looks like there is a threat to our well-being, it will step in to keep us from harm. Most of the time it will do that by using some kind of defensive coping strategy. Our neocortex is so good at what it does that we often don't even know we've shifted out of effective character-led strategy and into ineffective coping strategy!

Look at the image below. All of the horizontal lines appear slanted, right?

In truth, they are not. Each horizontal line is perfectly straight, but our brain is convincing us of an entirely different "truth." Even when you know for a fact that those lines are straight, your brain is still broadcasting the idea that they are slanted. It is very difficult to see them any other way. Unfortunately, the brain, while amazing and wonderful, is not always the best source of truth in our lives, as

Café wall illusion—R. L. Gregory and P. Heard, "Border Locking and the Café Wall Illusion" (*Perception*, 1979, vol. 8 issue 4, 365–80).
© Fibonacci / Wikimedia Commons / CC-BY-SA-3.0 / GFDL

we mentioned before. Sometimes that trying-to-be-helpful neocortex is making decisions to use ineffective coping strategies instead of above the line character strategies, in its effort to keep us from pain. Sometimes we have to find ways to dig for the *actual* truth—that lies in the heart.

Like most people, Eva was unaware that any of this was happening in her brain during the fated Monopoly game. And yet she felt *in her heart* that her response wasn't appropriate. She had hurt Sam, the exact opposite of her deepest intentions. Diving into the S+T=B formula helped her realize for the first time in her life where her negative competitive behavior (as opposed to achieving, which is positive competition) came from. Any situation in which there was a perceived "winner" and "loser," even a game with her son, triggered her desperate need to win to prove her worth and feel good about herself. And this was something she wanted to change.

As Mara likes to say, "I don't have to be a slave to my limbic system." We are not stuck with our ineffective templates, and we are not blocked from living out our best intentions. By identifying and understanding our triggers, templates, and truths, and the *context* of previous templates and current situations, we have the opportunity to make a different choice, in the moment and over the long-term. We can spend more of our time above the line, behaving in ways aligned with our best intentions, feeling happier with ourselves and the results of our behavior. That shift is key to our character growth, and like all important, lasting change, has to start in the heart.

Not long ago, we got an e-mail from Eva with a story about a tennis game. "We play, as a family, every so often," she wrote. *Uh-oh*, we thought. But this story surprised us. Eva had walked onto the court determined to make a different choice, to shift her heart, her thinking, and her behavior above the line, and to prioritize having fun with the family over winning.

Sure enough, ten minutes into the game her daughter, Jessie, hit a beautiful side-line drive, but it was out—at least Eva thought it was. "I think that was out," she said. "No, it was in," said Jessie.

EXERCISE: S+T=B INSIGHTS FOR TRANSFORMATION

Think of your worst "below the line" moment in the past six months. Now consider the following questions:

1. How would you describe your *behavior*? What about it was below the line?

2. What was the *situation*? Who was there? What led up to your moment? What was at stake?

3. What were you consciously *thinking* as you went below the line?

4. While resisting a blaming mind-set, can you identify the *trigger* or *template* that drove you below the line? What *needs to happen* for you to go below the line?

5. Is there something you believed absolutely in the moment? Was there a *truth* attached to how you were thinking and behaving?

6. What was your below the line behavior trying to achieve? What was its *purpose*?

7. Think of *context*. Can you think of another time when you behaved similarly, even if the situation was different? What can you learn about your templates when you compare them happening in varied situations?

8. What *one step* could you take differently the next time a similar situation or trigger pops up?

What happened next, Eva wrote, was familiar—the same old sensations of being triggered, the initial thoughts of arguing the point, the rise in her blood pressure, the feeling in her heart that she was *definitely right* and winning the point was critical. But this time, she was able to remind herself of the template that was in play for her and the void in her heart that it was trying to fill. And then she made a different choice than she had in the past. *It's only a game*, she thought.

"Okay, no worries," she said, and then smiled and prepared for Jessie's next serve.

Our hearts and minds work very well together to push us below the line into fear and pride or above the line into humility and love. Together, they explain the incredible range of human behavior. Denial is the enemy of growth, and truth sets us free to grow into who we were meant to be. When we put the clever brain and the beautiful heart together, we gain the insights we need for transformation.

CHAPTER 3

Voids, Wounds, and the Gold within the Heart

Ben was experiencing what should have been one of the most joyful times of adulthood. He and his wife were expecting their first baby. In the days after finding out, Ben was ecstatic. But as the weeks went by a new emotion emerged: doubt.

What if I'm just like my father?

Most of us at some point say, "Ha! I sounded just like my father/mother." But we often say it with a sense of love, respect, and humor, seeing all the ways that our parents were, in their own way, wonderful. Not Ben.

When we met Ben, what first struck us was his physical presence, with a body made strong by weekly hours in the gym. His gaze was direct and often challenging, as was his humor. Sarcasm was his go-to verbal position when he tried to connect with people, or disagreed, or didn't like the direction of a conversation, or felt uncomfortable. He rarely expressed an emotion without it.

We could see the pride instigating many of his behaviors, driving how he asserted himself in conversations in a way that was intimidating and diminishing. And as we got to know Ben, we learned about the fear that created the pride, and the source of his doubts about impending fatherhood. From the outside, his personality looked strong, but in fact he was incredibly fragile.

As a boy and into his early teens, Ben was a sweet, soft-hearted kid who was also a bit pudgy. He was often teased and bullied in school and had been hurt by the rejection and criticism. His father was a self-made man whose own parents had come to America the year before he was born. Ben's dad had grown up tough, and he had made his way in the world by staying that way—he believed. Soft-hearted Ben, his only child, was a mystery to him. Ben's father believed, out of a good intention to protect his son, it was his job to *make him tough* to prepare him for the world. He pushed Ben constantly, rarely showed warmth or expressed positive emotions, and even got on his wife's case for "making things worse" whenever she revealed her own softer side.

By high school, Ben had learned his father's lessons. He started working out constantly and, probably because of his adolescent growth spurts, lost all the extra pounds he was carrying. He became the class clown, the one who put others in their place before they could do the same to him. Out of fear of rejection, he closed off his heart, stopped expressing emotion, and turned to cynicism and physical strength to promote and protect himself. That's what he kept doing for the next twenty years of his life. Ben had developed a truth: "Being tough means being successful. If I'm not tough, I'll be rejected by the people I care about and bullied by everybody else."

He was very effective in his role, but as a colleague and leader, he was distant and intimidating. As the saying goes, "Hurt people hurt people." People were afraid of him, hurt by his remarks and actions. He had no close friends, only "mates" who also communicated through sarcastic interactions.

Falling in love with his wife had started to weaken his veneer of "tough strength," but during our work with Ben, it became clear that he was still blocking people from loving him. His relationships weren't as fulfilling as he wanted them to be; the idea of being vulnerable and authentic was too terrifying. Sadly, walls have two sides; he was also preventing himself from fully loving others.

The prospect of becoming a father made Ben realize one thing: this kind of life was exactly what he *didn't* want for his child. He had spent most of his life living out of the voids and wounds of his heart, and he didn't want to anymore.

Enter the Void—and Find Freedom

So what exactly are these voids and wounds? The voids are best seen as a *lack of*—usually a lack of love, esteem, self-worth, security, education, a parent, or achievement. The wounds come from *attacks on* our heart: rejection, criticism, being made fun of, failures we never moved past. Think about a time when you felt betrayed by a friend (if you're like most people, it probably happened at least once during the brutal middle school years!). Think about a lost relationship and how it ended. Think about a fight with a family member that went on for months. Think about Eva from the previous chapter, who was competitive with her son: she had equated love with praise as a child and so carried a template of a void of love because she had felt at times that her parents didn't love her as much as they loved her brothers. These are the types of common experiences that can grow the voids and wounds in our hearts, and they can impede the growth of our character. And they are all normal!

We try to fill those voids by promoting or proving ourselves with pride-driven behaviors, like pretending to know more about things than we really do, or inserting ourselves as the pack leader in a group, or speaking condescendingly about others to make ourselves feel good or better than others. We try to avoid the pain of those wounds with fear-driven behaviors, hiding from rejection and criticism. We may be overly nice to people, wanting their approval, or we may continually belittle our own views because we are not confident in ourselves. When we don't feel loved, or full of love, we compensate with ego or pride—self-promotion behaviors that mask the pain, like being overly competitive or perfectionistic. (Paradoxically, when we spend our lives self-promoting to retain our job, in a culture that values above the line behavior, we're at risk of losing our job.) When we don't feel full of inner strength and value (that comes from humility), we compensate with approval seeking, dependence, and avoidance. This is the X factor, and we'll explore it in more depth in chapter 6.

The past is not populated exclusively with bad moments, of course. There is much good in there, too. The wonderful positive experiences are captured in the core of the heart, strengthening your character, growing the gold within. By "gold," we mean all the forms of love—to be loved, to love others, to love yourself—and the humility to embrace that love. The good that emerges from that love—compassion and authenticity, joy, and dreams for yourself and those around you—makes you the fantastic human being you are.

Think about some of the best moments you've had with your brother or sister, parent, friend, or partner when you felt supported and loved. Think about a big, career-changing project that was your responsibility—and you and your colleagues or team nailed it.

Think about one of the first exciting moments with your spouse or significant other, the first time you realized you were in love with that person and that they loved you back. All these experiences help grow the gold in your heart.

You, like Ben and like every other human being, are exceptional at one thing: holding on to your experiences (we're sure you're also exceptional at all sorts of other things, too, but this one is universal). Throughout your life, from your childhood to this present moment, your experiences have had an impact on your heart, where you *feel*—they are your templates. It's common knowledge these days that our childhood shapes our life. In just one recent study, a team of researchers followed 710 Finnish families over seven years to investigate the consequences of early relationships for children's emotional development. The results showed that where early family relationships were healthy and supportive, children years later were better able to manage and respond to emotional experiences than those children whose early years were marked with problematic family relationships (such as authoritarian or disengaged behaviors).

We say childhood shapes our heart—the voids and wounds, and the gold within—and thus shapes our character. The wonderful gift we all have in life is to have the courage, knowledge, and wisdom to identify what's not effective and choose to strengthen our character and think differently to change behavior. You are more powerful than you often believe. When we surrender and humble ourselves to a higher power, we become even more powerful.

Panning for Gold

As kids, we were all asked to read aloud in class, sometimes while standing in front of the blackboard with all eyes on us. Depending on what kind of reader you were, you fell somewhere on the spectrum of completely comfortable to completely terrified. I (Stephen) was on the far end of fear, for good reason. My teachers didn't know something important about me, and I didn't either: I am dyslexic. When I was in the third grade, my teacher (probably trying to help me improve) constantly would ask me to stand in front of the class and read aloud. I would make mistake after mistake. And the class always laughed. Great entertainment for them, but total humiliation for me. In my heart, I developed a void, feeling a lack of self-esteem, and a wound, caused by the rejection of my peers every time they laughed at me. By the end of the third grade I had created a truth that I would never be good at school.

Not surprisingly, I eventually lost all interest in performing well. As I got older, I became a troublemaker, working to fill that void of esteem with rebellion, a counterfeit confidence that proved how clever and tough I was and how little I cared about authority or anybody in a position to judge me.

Third grade and my undiagnosed dyslexia was a bend in the river in my life, a key experience that shaped it. If you know much about how rivers flow and are shaped over time, you'll know that what accumulates in the bends of a river is silt. But if you've ever gone panning for gold, you'll also know that the best place to look for gold is not in the straights but in those bends. The gold *also* gets deposited along with the silt. We think of voids and wounds as the silt that collects around the gold in the heart. I certainly had my share of silt deposited in third grade, and that silt hid the gold that was also deposited. What is the gold in that bend? As a young adult, when I came to understand that I was perfectly capable of learning, I also discovered that I had a passion for it, a yearning to develop—myself and others. In fact, it became a deep part of my purpose, helping people along the sometimes-tough trail of learning about themselves so that they can grow their character and build their sense of worth. I've worked hard to replace the counterfeit confidence that showed up as being competitive and approval seeking with authentic and transforming thinking styles, as well as compassion for and a focus on developing others. I couldn't have done that without choosing to view my negative life experiences as growth opportunities, strengthening my heart, and creating a different heart attitude and mind-set.

I call it *turning your mess into your message.*

These experiences shape the heart, and we then live out of the four principles. When *situations* trigger us, those signals influence which templates are pulled up, and so influence our *thinking* and thus *behavior.* The voids and wounds of the heart impact our below the line behavior, spawning the coping strategies we rely on to get by in the haze of fear and pride. The gold within the core of the heart elevates us above the line.

The way we strengthen our hearts is by finding and expanding the gold within, and sometimes that means panning for it in the bends of the river of life—finding the positive character development in the tough experiences that may have caused our voids and wounds.* This is our "adversity quotient" (a phrase coined by author Paul Stoltz) because our character is developed through adversity—when we can find the good within it. We all have the ability to strengthen our character by choosing to incorporate the positive learning and growth opportunities within negative life experiences. It just takes a steady, gentle hand to wash away the silt and reveal the gold in the bottom of the pan.

We all have the potential to stay above the line under pressure or stress—by strengthening our hearts and growing our character. You're less likely to resort to below-the-line coping strategies when your boss calls you into his office, your teenager comes home with a lousy report card, or your sister-in-law gives you a backhanded compliment. Today, whenever Stephen has to read in front of a group, which happens in every workshop he runs, he goes to the core of his character to find the strength to prevent the triggers and templates of a vulnerable third-grade boy from ruling his life.

Rising above Your Voids and Wounds

We know a woman, Lia, who had a highly autocratic boss early in her career. He made her feel so belittled that even years later, she used to have a recurring nightmare that he was hired by her new company. She would wake up in a sweat and spend ten minutes convincing herself it wasn't real. What she began to discover through Heartstyles was that this early experience had created a wound in her heart that pulled her into easily offended behaviors. Any feedback from a manager *felt* like an attack. So Lia made a promise to herself that we refer to as an *inner vow*: "I will never trust my boss again." Inner vows only add more silt by reinforcing our below the line coping strategies. Now, many of us have had overbearing bosses—hopefully not ones that cause recurring nightmares!—but how many of us have really taken the time to grow away from that experience by finding the good within it?

We don't believe in *blaming* bad behavior on our past experiences or the people in our lives; blame is just a fear and pride coping strategy. However, we do believe that making better choices and developing our character requires us to *understand* the roots of our behavior. To discover your authentic self and live (mostly!) above the line is a journey of understanding the gold within and the voids and wounds that shaped our lives.

We can choose to allow those voids and wounds to be the fuel to make a difference in our own and other people's lives by seeing them as growth opportunities. Some people can rise above their voids and wounds, accepting them as part of their character journey and making them part of their fuel for living above the line in courage, humility, and love.

Life is so worthwhile! No matter what our background, what might have happened to us, we all deserve to be happy. Negative things can happen to us, and they can shape and mold us, but they don't need to define us or destroy our life. In the end, what defines us is how we respond to life, and the decisions we make about who we are and who we want to be. *We* are the captains of managing who we become.

If we are living out of the voids and wounds in our hearts, changing our below the line behavior will be difficult because they will keep pulling us down. The gold in your heart might be buried by the silt of voids and wounds, yet it is key to transformation.

As the birth of his first child approached, Ben chose to go on that journey. He had both voids and wounds that played out in his behavior. When he began asking these questions—"What happened to my heart? What shaped my life?"—Ben's truth was being softened by love.

Ben today is quite different from the Ben we first met. He's a loving dad to his children and a fun guy to be around. He can still be cynical and sarcastic at times, but he's aware of it and no longer uses sarcasm to belittle people. Instead he has learned to keep his banter fun and about things, not people. He is always working to strengthen his character. Ben is applying the wisdom well captured by Maya Angelou when she wrote to her daughter, "You may not control all the events that happen to you, but you can decide not to be reduced by them."

In the next exercise, we share a technique to help uncover the root of your behavior. Ben, like all the people we've worked with, used it to understand his thinking and behavior more deeply. We hope it helps you do the same.

The Cement That Sets Like Stone: Identifying Your Inner Vows

A little girl, of about five years old, is watching her mother prepare for a dinner party. The table is covered with the nicest linen. Her mother has put on her fancy dress. The girl feels the excitement and

EXERCISE: A TEMPLATE TIMELINE

In any seven-year period, most of us experience significant change in our lives. The differences between seven, fourteen, twenty-one, twenty-eight, and thirty-five are often dramatic, as are the experiences we accumulate along the way, both good and bad. Those experiences from each "season" in our lives shape our hearts, form templates, and influence our character and our behavior.

Identifying them is the key to understanding the root of our behavior. Take the time now to give yourself this gift. It may be one of the most important things you can do to discover the things that have held you back, and how you can be released to shift your heart, your thinking, and your behavior.

Create a grid on a piece of paper that looks like this one:

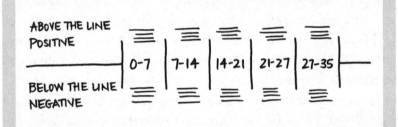

I. Above the line, write down one or two vivid, positive, joyful memories that pop up for you when you think about each season of your life at that age range. Who had a profound positive impact on your life? How? Make a note of the gold from those experiences that still lives in the core of your heart.

2. Next is the part that requires courage and humility. Take a moment to remember why you're doing this, and what it is you want your life, your relationships to be. Think about being enlightened, finding your authentic self, having insight for transformation. Now turn your mind and heart to the less positive experiences you remember in each season. What does each make you feel in your heart as you think about it? Is there *a lack of* or *attack on* your heart from the experience?

3.   Now, based on what you've uncovered, what are some of the templates, above the line and below, that may be leading to your effective and ineffective patterns of behavior? What truths do you hold about how life works that could be helping you or limiting you? What "truths" might actually be lies you've told yourself for years? In looking at these experiences, when do you think you are triggered, and how?

4.   Finally, look over all your notes. How are these voids, wounds, and gold, these triggers, templates, and truths shaping your life today? Write down what you are committed to stop living by, start living by, and continue living by to be your best self. Apply these daily until they become the authentic you.

wants to be a part of the special night; she asks over and over if she can help take the "special" glasses over to the table. The mother finally relents and gives the girl a tray to carry three of the glasses into the dining room. The glasses were inherited by the mother from her mother, who had inherited them from her parents. "Be careful," the mother says. "They are Mommy's special glasses. Don't drop them."

The little girl stares at the glasses as she's walking, making sure they don't topple. Just as she's getting to the table, she misses the edge of the rug, stumbles, and crashes to the floor, along with the tray. And the glasses.

"Get out of the way! Look what you've done!" the mother shouts, grabbing the girl out of the fine shards of broken glass that are scattered everywhere. "I told you to be careful!" the mother yells. The girl stands nearby, crying and watching her mother carefully begin to pick up the small pieces.

We've all been there. Perhaps as the child, or as the parent, or both if we're honest! That mother was stressed, and mostly frightened for her daughter, but a five-year-old couldn't see that. From the little girl's perspective, she had tried to help, she had made a mistake, and she had made her mother angry. Out of that experience, and other similar ones in trying to please her mom, she developed a void of self-worth (I'm not good enough), and an inner truth: Anger = not loved. If you want to be loved and respected, don't make mistakes.

That little girl was me (Mara). Today I can tell you that I am a recovering perfectionist! I used to get so angry about typos in reports that went to clients. There were so many moments when, driven by fear of criticism and by pride to prove my worth and reputation, I became stressed, abrupt, negative, and critical. While I'm usually a well-liked leader, I could turn into a perfection monger who sucked the fun and joy out of anything within a fifty-yard radius! In those moments I was fulfilling an *inner vow*: I will never make a mistake again. I *have* to do it right.

Every one of us has made an inner vow at some point in life. An inner vow is a promise we make to ourselves within our hearts, usually designed to fill a void or protect us from a wound. People frequently *talk about* their inner vows without realizing they *are* inner vows and may be limiting the choices they make in life. They are usually *I will never*, or *I promised myself*, or *Take my advice, never* . . . kinds of statements. You may have said to yourself or out loud to somebody else any of the following, or something similar: *I'm never going to work for a big company again. I promised myself I would not date somebody who has been divorced. I'm never speaking to that person again. I'll never trust a man/woman again.* Sometimes these inner vows aren't strong and they don't last. Sometimes they last a lifetime until you admit it. What's most important is that they close us off to possibilities and growth and create a stuck mind-set lifestyle.

Sometimes they *can* push our hearts into humility or love. For instance, we have met many people like Ben who worked to be loving and supportive parents specifically because their own parents had been the opposite. They had made an inner vow not to treat their children the way they had been treated, and used it positively to grow gold out of the fear. Another example of a positive inner vow is not doing conflict on e-mail!

More often, though, our inner vows reinforce our below the line coping strategies. They prevent us from seeing the truth of a situation or a person. They limit us—our relationships, our opportunities, our self-awareness, and our growth. Mara's inner vow of *I will never make a mistake again* sent her down the path of perfectionism, which was stressful—for her, but *more so* for those around her being sucked into the perfection vortex!

Our inner vows and our truths are frequently linked. For Mara, they certainly were. In her early thirties, Mara began to examine what had shaped her heart, and how it was still shaping her thinking, her behavior, and her life. She was able to address the void, the vow, and the truth and grow her character. But, as she says, "I had to have a huge wake-up call of my own mortality to make me stop and really face myself. I got cancer, and it was severe. I survived, thanks to medical technology. But I really believe that the decisions I made to truly look at who I was and how I was living my life helped me fight with love, not fear. It wasn't easy, and it took a conscious decision to let go of the outcome and focus on being the best me I could be, no matter the end."

Mara's perfectionism, like many below the line behaviors, seemed to benefit her enough that it took her years to understand the negative impact. Our inner vows can feel like a form of having control, but sometimes they can cripple us. Holding a grudge is willful retention of an inner vow. It sends us into ego-driven pride—it can set like cement and keep our heart below the line in negative emotions. Inner vows, although seemingly useful, are still limiting—sometimes seriously and sometimes not so much. Stephen recalls, "The first time I went skiing I had a terrible time and I felt like a failure. Out of pride, I vowed, *I will never go skiing again*. It took me ten years to work out that I just didn't have the help I needed that day to learn, as a beginner. Finally, I said yes to a ski trip with friends, took some lessons, and learned to love skiing. My inner vow didn't limit my life dramatically, but it kept me from something I certainly would have enjoyed for those ten years."

EXERCISE: TAKE YOUR VOWS (AND LET GO!)

1. List a few promises or vows you have made to yourself. If you're struggling to come up with some, think about a time that you've said one of the following in relation to some aspect of or experience in your life:

    "I will never . . ."

    "I will always . . ."

    "I promised myself . . ."

    "I hate . . ."

    "I must . . ."

2. Now consider the following questions for each of them.

    What was the vow actually doing for you? How was it protecting you or helping you prove yourself?

    In what ways can that vow have limited your life or the choices you've made?

    What positive promise can you replace it with that will help build your life?

3. Not all our inner vows are obvious to us. Think back to a few decisions you've made over the past year or two, such as whether or not to take a job or pursue a relationship or take a trip. If you dig into that decision, can you find an inner vow that may have influenced it?

4. Now, taking your courage in your hands, ask yourself, "Who do I need to forgive?" and "What needs to happen for me to decide to let this go?"

Six Keys to Growing your Character

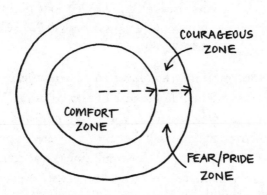

Growing your character is about having the courage to expand your personal comfort zone within yourself. Staying in our comfort zone where we feel safe, secure, and in control can in fact be the deception of living in self-limiting fear and ego-driven pride. In nearly every aspect of life it takes character for you to be your best self. Interestingly, when we achieve more in life, it presents us with the opportunity to grow or stay at the edge of our comfort zone. For example, you do well at work and you get a promotion opportunity. Immediately you can experience fear of failure, fear of what others may think of you, questioning whether you can do the job, even feeling you will have to use a *fake-it-until-I-make-it* strategy. Like it or not, growing character takes courage to face our fears and not mask them with ego-driven pride or hide from them. Developing ourselves means seeing this as the *courageous zone*—to learn and grow.

1. Be courageous with myself

- **Work on your self-awareness:** Get to know yourself more deeply. Create an S+T=B journal. Question yourself on your reactions: "What situation/s triggered me today? Why did I react that way? What was happening for me?" Set aside five to ten minutes every day to journal (in a book or on your device) what you discover. Look over your notes at the end of every week. What conclusions do you draw? What can you practice differently next week?

- **Find the emotions that have ruled you:** Recognize when you go below the line because of fear or pride. Journal any triggers, templates, truths, voids, wounds, and inner vows you uncover. Set yourself a goal every week to practice self-control to chip away at those emotional reactions—S+T=B! When are they most likely to get triggered? What courage do I need to find for the next time it happens?

- **Find the gold in the bends of the river of life:** What is one thing, even if it seemed difficult initially, that was a good outcome from a negative past experience?

- **Do something nice without others knowing:** Set a goal of one kind gesture once a week, or every two weeks, whatever works for you. Do what you can to not get noticed! It's amazing how that builds character and lessens our need for approval. Only you know, and that's what counts.

2. Be courageous in new activities

Take a "calculated risk" and decide on doing something new outside of your usual environment. Do something or learn something new: cooking, dance, music, or language classes; public speaking; or gym classes, or get a lesson so that you can then use the equipment on your own. You could even stretch it to something physical you've never done, which helps build character: skiing, abseiling, swimming in the ocean, paddleboarding, parachuting, or rock climbing. Having a goal to walk halfway up a mountain is huge for someone terrified of heights. We had dinner recently with a leader celebrating his personal success—two years ago we were with him on that same mountain when he turned around, paralyzed by fear. He went back. He stretched his character. He achieved.

3. Be courageous with others

- **Allow yourself to not feel you have to act in certain ways.** Practice giving one honest reaction a day. Connect with how that feels in yourself, and make note of what happens for others each time you are authentic.

- **Admit mistakes when pride has got you, and choose to make things right.** Do what you can to make amends for your actions. Depending on the situation, that might mean you need to apologize. Be quick to admit when someone has a better idea or you could have done things better.

- **Honor other people when they are ineffective.** Choose to see someone else positively. You don't have to condone ineffective behavior, but you can choose to not condemn the person. Look past the behavior and consider what might be happening for that person.

4. Practice forgiveness

Our below the line behaviors rarely make us feel good about ourselves. We can carry shame and guilt—for how we've treated people and poor choices we've made—that push us further into denial and make it harder for us to release the gold. Make a point of acknowledging these feelings and forgiving yourself and others. There's a wonderful part of our humanity where we *want* to and *need* to forgive because it releases us from the cement of judgment. Life is not perfect, and people are not perfect—that's why forgiveness exists on our planet. The power of forgiveness is one of the most releasing, life-giving practices one can live by.

5. Practice gratitude

You can develop gratitude by intentionally making it part of your everyday life. Challenge yourself to one hundred days of gratitude! For example, you can end each day by thinking of one, two, or three things that you are grateful for and note them in your journal. Or keep a gratitude jar at work or home: write each thing you're grateful for that day on a small piece of paper, date it, and pop it in the jar. Every month, look into the jar and pull a few things out. These are great reminders of so much we all can be grateful for.

6. Practice meditation or prayer

Have some quiet time every day—even five minutes. Use an app or music first thing in your morning, as you exercise, or as you travel to and from work. Maybe focus on gratitude and what you're thankful for during this time as well.

Lasting transformation comes from expanding the gold within our heart, to fill the voids and heal the wounds. By strengthening our heart and growing our character we can change the shape of our lives, now and in the future. But to do so, we have the opportunity to release ourselves from some of our more negative past experiences, find the gold, and shift our heart attitudes for good.

Personal development is the awareness of how the four universal principles, triggers, templates, truths, voids, wounds, and vows positively and negatively impacted our lives. We encourage you to gently, kindly wash away the silt (those negative experiences) as you would if you went gold panning—and find those specks and nuggets of gold that have been deposited in your life and your heart, even through tough circumstances, and release yourself to become your best self!

CHAPTER 4

The 16 Common Behavior Styles

Before you read this chapter, we suggest you go online and complete your *free* Heartstyles Indicator Self-Score. You can do this by going to the appendix section and using the QR code to direct you to the website.* You can then enjoy reading this chapter in the context of knowing your own results. It can be read from start to finish, or simply used as a reference for getting to know what certain above the line and below the line styles feel and look like. There's also further explanation in your Personal Development Guide (PDG for short), which you can download upon completing the online HSi. Understanding the 16 styles will help you put together the first three chapters and what you've already learned about thinking and behavior.

Morgan stood in the kitchen watching her mother furiously scrub the counter around the stove. She couldn't see the phantom spots or streaks that seemed to be driving her mother over the edge.

Nancy looked up from her work and called into the dining room, "Not *those* placemats, John! The *nice* ones." Morgan's dad lifted one of the placemats Nancy had bought two months ago off the table and left to go find the "nice ones." Morgan caught the slightest eye roll and watched him silently replace the placemats.

She remembered what it had been like to be told to help when she was young but then watch as her mother redid everything, snapping at her that things weren't done right.

"What can I do to help, Mom?"

"Nothing. It's all taken care of. Just go spend time with your brother." The clipped, abrupt speech sent Morgan fleeing from the kitchen.

Morgan knew her mother had been cleaning and cooking all day, trying to make everything "perfect"—and that she was tired and edgy.

"Is the cleaning frenzy still underway?" her brother, Michael, asked as she sat on the couch.

"Of course."

"And you're not helping?" The sarcasm, always there, was deeper tonight, harsher. Why was *she* getting the brunt of it?

"You know she won't let me. And why aren't you in there?"

"I'm not going into the lion's den." Another smirk.

His attitude was grating. "Right, of course not, nothing that puts you in danger of actually having to *deal* with her."

"What's that supposed to mean?"

"You haven't shown up for dinner in months!"

Michael got up and stormed out of the room.

Morgan dropped her head in her hands for a few minutes and then took a deep breath and slowly made her way back into the kitchen. "Everything looks really nice, Mom."

"Well . . . thank you."

"Did you get new drapes in the living room?"

"Yes. I must have looked at a hundred different choices. Of course, your dad was no help." The chatter about the endless search for drapes turned into a story about the hunt for just the *right* hand towels. When they finally sat down for dinner, Michael would hardly speak to her. Her dad chatted about his improving golf game and complimented the food. Her mother complained that the meal hadn't turned out the way she wanted. And then it was over.

On the drive home, Morgan replayed one scene after another.

Why couldn't her mom just relax and enjoy the family dinners? Why did her dad just put up with her obsessive perfectionism? And why was her brother avoiding them all, or being snide and hurtful when he showed up? She had been looking forward to seeing him, but then she'd attacked him for not being around. Why had she done that? And why couldn't she talk to her mom about how her intensity was making them all feel? It was as if her mother thought that if she could just make the house perfect, everybody would have a perfect time. She couldn't see that the opposite was true, and Morgan could never find a way to tell her.

Morgan turned into her driveway and switched off the engine. With a deep sigh, she leaned her head on the steering wheel for a moment. She loved her family, and she just wanted to have happy times with them. Why was it so hard?

Let's face it, we are all exceptional at analyzing other people's behavior! But how good are we at spotting the behaviors that are keeping *us* from our best intentions and those that are freeing us to live our lives feeling happy, fulfilled, and loved? To help people understand not only why we do the things we do, but also *what* we do that is either effective or ineffective when we are triggered, the two of us spent more than a decade researching behavioral traits, assessments, philosophy, and principles.

We wanted to give people the language and a framework for identifying and talking about the things they intuitively understood. We also wanted to develop a tool to help people grow in self-awareness, because we know how effectively fear and pride work together to create denial. Denial, we have found, destroys self-awareness.

The four principles—humility, love, pride, and fear—have four behavior styles each that emerge when we are triggered in certain situations. The model connects the four principles that live within the heart with the cognitive processes of the brain to describe the B in S+T=B (situation + thinking = behavior).

That model is the basis for our Heartstyles Indicator,* an online survey that reveals the behaviors that are more or less dominant in our lives right now. It is not another type of personality indicator, a static label of who we are, but a dynamic *life indicator,* a *character-development tool* describing how we are living our lives today. From one year to the next, from one month to the next, the results can change as we grow and strengthen our character and work to live our lives more above the line.

Each of the 16 behavior styles are not "good" or "bad"; they are all human and normal. We each are strong in certain above the line behaviors or resort to certain below the line coping strategies based on our unique triggers, templates, and truths, and the voids, wounds, vows, and gold of our heart. These behaviors have developed over time as we learn to cope, to get ahead, to protect ourselves, to prove ourselves. When we learn more about them, we're able to spot them and trace them back to what is happening in our hearts and minds. And as we interact with others, we learn to understand their behavior better and to have compassion for the Nancys of this world, because we are all facing similar struggles. We begin to see *why* the below the line styles aren't effective at getting us what we really want, and we have the understanding we need to shift above the line into more effective behaviors. And when we can do that, we live our lives with more successes, better relationships, and greater purpose.

When we spent time with Morgan on a program, she asked the question: "What shaped my mother's life, her templates, her inner vows?" Morgan knew her mom had been raised by high achievers and had no doubt wanted to live up to those expectations. Morgan's empathy rose for her mom, and she decided to stop judging her, accept her for who she was, and stop taking her comments and behavior to heart. A year later, Morgan said to us, "You can't believe how much my mom has changed!" Love creates a safe place: in setting boundaries and not taking her mom's behavior personally, but seeing it for the coping strategy that it was, Morgan was able to have compassion and a less-stressed attitude. Her mom picked up on the change. Slowly, Nancy stopped trying to make every meal perfect and spent more time focusing on the quality of the time together. She asked questions of Morgan and her brother, Michael, that were genuinely loving rather than criticizing. Michael started to feel safer and so his behavior changed and softened a bit, too. Eventually, that family started enjoying being together. Family dinners were no longer a danger zone!

It's never too late to be happy. Did Nancy go on a personal development journey herself? No. She didn't really have the language to identify what had changed. But Morgan, and later Michael, did. They decided to rise above the line, and their humility, courage, and love helped Nancy feel loved and accepted, and not needing to prove herself, which helped her change, too.

The 8 Above the Line Styles— Effective Behavior

Take a moment to think about a friend or family member you truly admire and trust, the best boss you've ever had, or maybe a public figure who meets their commitments with integrity. Now think about a moment in your own life that you feel you handled well. We're sure you thought of the person and the personal moment you did for good reasons. What are they? What behaviors or qualities do you associate with the people you admire, or the moments in your life when you felt most effective?

When we do this activity with groups of people, we see the same lists of words come back—words like authentic, reliable, courageous, caring, supportive, compassionate—regardless of what kind of a group we're working with or where in the world we are. Effective behavior is effective behavior regardless of circumstance or culture or position because it's driven by humility and love. We aspire to it and admire it because we see how it helps people achieve their highest goals, and because we recognize the inner strength we feel and our positive effect on the world around us when we are above the line.

Before we get there, this chapter you're reading now allows you to spend time getting to know these behaviors and learning how to spot them. As you read the following pages, think of what these behaviors can look like in very different scenarios in your own life. Ask yourself, *When have I been strong in that behavior? Where was my heart? How did it work for me?*

The Four Courageous Humility Behaviors

When the heart is operating out of humility, it manifests behaviors of personal growth like courage, diligence, honesty, and learning, and purpose-driven results. Behaviors rooted in the humility quadrant are authentic, transforming, reliable, and achieving.

Authentic behavior is driven by a heart attitude of being one's real self. When we are strong in authentic behavior, we live with courage and integrity, doing our best to stay true to our values. We can be honest with ourselves and transparent with others. We have the kind of humble confidence that comes from a sense of inner worth, so we can admit to mistakes, laugh at ourselves, and love the skin we're in. People who are high in authentic behavior radiate a calm command under pressure and carry a sense of purpose.

Think about a time when you took the harder road or made the tougher choice because you knew it was the right thing to do or was most aligned with your values. Consider a difficult conversation you had in which you were able to be straightforward about your thoughts, feelings, and opinions without blaming or criticizing or having to prove how right you were. These are moments of authenticity. That's what Morgan, who we wrote about at the start of the chapter, was longing for in her relationship with her mother. Neither of them had the courage, yet, to be honest and straightforward, and courage is an essential element of all the humility-driven behaviors.

Transforming behavior emerges from a heart attitude of personal learning and development. It is passionate about and committed to growing in wisdom and maturity, being vulnerable and teachable. We are open to opportunities to learn from any person or experience. We're unafraid to try new things or new ways of accomplishing goals or to admit that we don't know or need help. People high in transforming behaviors tend to be aware of their strengths and weaknesses and don't shy away from receiving feedback. They are at peace with the journey they are on.

Years ago, Stephen was speaking at a conference. He was one of the lesser-known presenters, an "opening act" for some rather famous people. One of Australia's most successful businessmen was a part of that group. Stephen met him during the conference and after a few minutes of chatting, the man offered to continue the relationship by visiting Stephen at our offices when he was back in Sydney. He wanted to learn how we operated. Of course, Stephen thought it should have been the other way around, and wasn't sure this high-flyer would follow through. A month or so later, though, he plonked himself down in Stephen's office, asking question after question, showing his deep interest in other people's experiences and ideas. He took notes as they talked, creating a list of actions he might take based on what he was learning. He also shared a load of valuable insights and was transparent about missteps he had made along his path. It was no wonder this man had become so successful and built an amazingly effective culture in his business. His humility and his own passion for learning and growing was so inspiring and created a safe place for others to learn and develop.

Reliable behavior is driven by a heart attitude of being dependable, diligent, and conscientious. We value discipline, consistency, and keeping promises or honoring commitments—without rejecting fun and flexibility. To maintain these behaviors and live these values, we learn when to say no, when to delegate, and how to avoid overcommitting. We meet deadlines and deliver what we say we will deliver. If we can't, we are transparent about why, without resorting to blame or excuses. For people who score high in reliable behavior, their word is their bond, they honor other people's time, and they radiate trustworthiness. They have effective systems to keep them organized (follow-up, forward planning, and calendar appointments); this may sound easy and basic, but it makes them dependable.

Imagine you're on a flight across the Atlantic Ocean. How would you feel if the pilot had chosen not to bother with all the points on the pre-flight checklist? We all expect people in certain professions to show reliable behavior—pilots, accountants, doctors. But isn't it wonderful when we can also count on colleagues and friends? Isn't it easier to make positive things happen together? And how much do we love to hear, "I know I can count on you"? In our chaotic world, it's some of the highest praise we can offer or receive. Being reliable increases trust, and being unreliable decreases trust.

Achieving behavior is all about getting things done with excellence—delivering our personal best, not perfection. When we exhibit achieving behavior, we are operating from a heart attitude of achieving goals, getting results, making a difference, and working toward a purpose beyond self-interest. We plan ahead and go the extra mile. People high in achieving enjoy winning at work and in life with "healthy competition"—doing so with honor, integrity, and humility.

It's easy to notice high achievers working on a flight, for example. Airline cabin crew don't get tips, so their internal motivation to be a high achiever is obvious to us the customer by their service and attitude. High achievers don't wait to be asked; they *look for things to get into, not out of.*

The Four Growth-Driven Love Behaviors

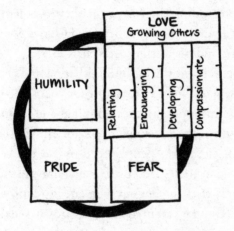

When the heart is operating out of love, it manifests honor, respect, loyalty, kindness, and assurance. Behaviors rooted in the love quadrant are relating, encouraging, developing, and compassionate.

Relating behavior comes from a heart attitude of building meaningful relationships. We work to understand others better and have respect for their perspectives, regardless of how they may differ from our own. We get to know people by showing an interest in them, relying on good social skills, and being active listeners. We are respectful and polite. We remember people's names and make them feel comfortable and cared for. Being high in relating behaviors allows us to connect in a genuine way with people from all walks of life. People high in relating behaviors radiate kindness, graciousness, and genuine interest in others. It's being relational instead of transactional in our interactions.

"What's routine for you is relational to me." Customers want to receive heartfelt service that feels relational. For any person serving customers, the culture of heartfelt service can die off and it becomes routine, to the point that the customer feels the server does not care. Our son Nathan has his own business in the automotive industry. It's a tough industry; most customers only see Nathan when something is wrong with their vehicle! He often sees the same problem day after day—that is routine for him. But he makes sure he keeps his passion fresh and stays relational with the customer, understanding their need for their problem to be solved and to be treated with care and expertise. He goes out of his way to serve with heart, and his business has thrived.

Encouraging behavior helps build other people's courage, character, and self-worth. With a heart attitude focused less on self, we take satisfaction in supporting, recognizing, and motivating others to build their inner character. We look for things that have been done well or for signs that people need praise and recognition in tough moments and then give positive encouragement. An important point here is that the encouragement is genuine, not false—praise where it's not warranted is an approval-seeking behavior and others can usually see through it. People who are high in encouraging behavior will often praise frequently, even if for small things. They know that positive, loving support helps build inner strength and contributes to the gold within. We like to spell it "in-courage-ment"—we are meant to instill *courage* into hearts of people, from children and friends and loved ones to staff to bosses.

Stephen is a passionate rock climber (he'll tell you it was in the fine print of our marriage contract that Mara somehow didn't see! "Thou shalt hang off a rope with thy husband . . ."). The importance of encouraging behavior is never more crystal clear for us than when we're clinging to a rock face. Encouragement is an essential part of climbing—partners check in with each other frequently, praise each other on tough moves, and cheer each other when they reach the next ledge or summit. It quiets the nerves, calms the fear, and brings a sense of belief in self that is necessary to be successful—in the sport *and* in life. Notice how sporting teams know how to encourage each other even after a mistake. People high in encouraging behavior recognize that we all need frequent encouragement, but we especially need it in the toughest times.

Developing behavior is driven by a heart attitude focused on other people's best interests, making time to serve them so they can reach their potential. We effectively coach others into personal growth through constructive feedback, and are prepared to be open and honest to build the character and skills of others without making them feel lesser as a person. We refer to it as "care-frontation" rather than confrontation, combining objectivity and empathy. People who are high in developing behaviors don't shy away from authentic, honest conversations about the need for improvement or growth, but they radiate a belief in the inner worth of others and a true desire to help them. And that makes them incredibly effective in coaching and leading others.

Martin was passionate about mathematics, so of course he would want to help his daughter when she was struggling with her math studies in her final year of high school. But every time they reviewed her calculus homework, it ended with Kelly snapping at her dad, still not really understanding the concepts. Martin was baffled. Surely math skills were hereditary! It was only after some loving intervention from Martin's wife, Jane, that he realized he needed to ask Kelly how *she* thought he could best help her, rather than deciding what was best for her based on *his* perspective.

They set up a strategy that Kelly felt would work, enabling Martin to help her in a way that made her feel capable, not inadequate. Months later, Kelly ended school with excellent marks. After leaving school she even gave her time to tutor other kids in math, helping them in the same way her dad had helped her.

Every one of us has the opportunity to help those around us be their best selves—as a parent, a partner, a leader, even a friend. It doesn't mean finding fault, criticizing, or correcting in a controlling way. It means helping people recognize their development opportunities. It can even mean exiting somebody from a team or organization if they aren't the right fit. It's hard to develop, though, because we are often so afraid of offending others, especially people we care about. And when others are driven by fear, even well-delivered constructive feedback can feel like an attack.

We have observed leaders using a variety of styles to give feedback and coach people in their development. Some live by "drink the soup while it's hot"—addressing issues immediately while they're still current. Others wait for an MOT (moment of truth)—discerning when the time is right to discuss the issue. There is no "one size fits all." The key is love and compassion (the next style).

Compassionate behavior emerges from a heart attitude focused on seeking to understand others, their behavior, and their circumstances. When we are being truly compassionate, we look past surface behavior and into the heart of others to understand the deeper "why," without judgment. We can ask, "What might be the triggers, templates, truths, or the voids, wounds, vows, and also the gold within?" We make the effort to not just react, and instead consider the past and what shaped a person's life, to look at how the situation might be triggering them, and to examine their positive intentions. Consider S+T=B. What situations are triggering the behavior?

People sometimes confuse compassion with condoning. Being compassionate *does not mean condoning or agreeing with the behavior* or ignoring it. It simply means choosing to be objective and nonjudgmental. It allows us to understand the source of people's behavior so that we can take it into account as we decide how to respond in an above the line way, possibly into authentic or developing behaviors.

We often say that the "bookends" of above the line behavior are authentic and compassionate because together they help us value our own inner worth and understand the inner workings of others, and that makes us powerful, positive forces in the world.

The Eight Below the Line Styles—
Ineffective Behavior

Not long ago, we had a rare experience. One participant received the results of his Heartstyles Indicator and just didn't believe them. This man had taken a 360-degree version of the Indicator, which meant that his report included averaged responses from six of his peers. He was upset by what he was seeing in those scores: higher results on below the line behaviors, lower results on above the line behaviors. His own personal assessment looked very different, with opposite results on those same behaviors. Sometimes those who struggle most with their HSi results can be struggling the most in life. Fear can lead to pride, which leads to denial—he appeared to have no idea how his behavior was affecting the people around him and his ability to achieve what he wanted in life. He needed more proof to overcome the denial he was facing. If someone is genuinely not comfortable with their results, we offer another Indicator for free. "Okay," we told him, "choose six other people and we'll send them the survey."

When the averaged results were ready, we took him aside to share them. They were almost identical to the results from the original six responders. "I never knew," was all he could say. "This is so humbling, and I really want to discover how I can improve myself." That was the start for his breakthrough.

We've already learned that our brains are not always the best indicators of the "true truth." The negative emotions that often come with our below the line behavior—guilt, self-doubt, shame—feed a downward spiral that can keep on dragging us below the line.

The most effective way to decrease ineffective behavior starts with one simple step: admit it exists—break the blindness. When you can recognize that you're using a below the line behavior, you have an opportunity to make a different choice. To shift above the line, you need to have compassion for yourself, and to acknowledge that all of us resort to these coping strategies to protect ourselves when we're driven by fear or pride. We may resort to them when we are triggered by other people's below the line behaviors. We may resort to them when we're living or working in a culture or with people who are below the line.

As you read on, keep reminding yourself that the below the line behaviors you'll be learning to identify are normal. Every one of us has resorted to these behaviors at some point—probably in the past month! The two of us have worked with these ideas for decades, and we still slip below the line.

It's especially important to realize that *most ineffective behaviors are born out of a good intention* that is then overridden when we're driven by fear or pride. It's like this: your eleven-year-old daughter comes into the room, ready to go out with her friends to the fun fair. As the parent, your first words could be, "Hello, darling, how lovely you're going out with the girls. Have fun! I love you!" But instead, they are, "Where's your cardigan? It's cold outside and you don't want to catch a cold!" Good intention = make sure child is warmly dressed on a cold day. Fear = if she's not dressed warmly, she'll get sick, which means I am a totally hopeless parent. Behavior experienced by child = annoying parent treating me like I don't know it's cold.

This scenario happened to Mara and our daughter, Tamara, when Tam was eleven years old. She had the grace to say calmly to Mara, "You know, you're just like your mother" (big gasp from Mara as her hand goes to clutch her heart—noooo!!!). "You're an overprotector. I have my cardigan in my bag." Mara had the grace to see what she'd done and apologized to Tam. "First, what psychologist have you been hanging around that you even know what that means?! And second, you're right! I'm sorry!" We still laugh about it twenty years later!

Each of the eight above the line behaviors has a "counterfeit" below the line behavior, even though it is born out of a good intention. Striving is counterfeit excellence, while achieving is genuine excellence; competitive is counterfeit improvement, while trans-forming is genuine improvement.

Think of each of your below the line tendencies as a "friend" who can introduce you to your "best friend." Go in with an open mind and an open heart and you'll come out with powerful insights for being more effective and just plain happier.

The Four Ego-Driven Pride Behaviors

When the heart is operating out of pride, it manifests self-promoting behaviors like being condescending, perfectionism, and winning at all costs, which block us from really connecting with others. Behaviors rooted in the pride quadrant are sarcastic, competitive, controlling, and striving.

Sarcastic behavior is driven by a heart attitude of needing to be the smartest, wittiest, or funniest person in the room to build a sense of worth. The good intention behind sarcastic behavior is often that we are trying to build relationships with others, but we don't have a strong enough sense of authenticity or inner strength to be sincere. (In fact, sarcastic is the counterfeit of authentic, often disguising itself as straight-talking frankness and confidence.) We need to keep a safe distance from the emotions that make us feel awkward or vulnerable in those interactions. So we revert to cynicism, sarcasm, and quick-witted comments to relate and connect. The trouble is, a lot of times, these can be intimidating or even hurtful and can become "intellectual sport."

Morgan's brother, Michael, whom we met at the start of this chapter, is a perfect example. His family's dynamic makes him uncomfortable. He uses sarcasm to distance himself from the situation, to imply that it doesn't affect him, but of course, it does. Driven by pride, which grew out of fear of his mother's criticism, he tries to prove that he's above it all, that he can't be wounded. In the process, his cutting comments wound others. But the heart of his behavior is that he's trying to communicate that he's uncomfortable—that he wants the same thing as everybody else, which is a fun, loving family.

Sarcasm is the basis for so much of our humor. So what's really wrong with it? Sarcasm is a *counterfeit authentic*—we convince ourselves we are being real, but we couch the truth in sarcasm, thinking it will be less painful. Trouble is, as it's gone below the line, its effect will *not* be positive. It can be self-deprecating humor that you think will make people feel comfortable, but you're just shrinking yourself, and that's not being authentic to yourself. It's getting people to laugh *with* you *at* somebody or something else—making a joke instead of really saying what you mean. It can be hostility disguised as humor: you're communicating a negative opinion without owning up to it.

Good-natured banter and true joking around is playful, fun, friendly. Wisdom is paying attention to where your heart is in the moment: knowing the difference—whether you're being driven by pride or by love—and knowing when to hold back on that sarcastic remark, no matter how funny you think it may be. Sometimes we need to be honest and admit it's really our pride showing off how funny or clever we think we are.

Competitive behavior is based in a good intention of wanting to be the best we can be, to win or be better than others. But when pride is in the driver's seat, the need to win at all costs can lead to selfish ambition and a habit of constantly comparing ourselves to others to make sure we are better. Overcompetitive becomes over-comparing becomes constricting and never good enough. It can push us to exaggerate, manipulate, and even cheat. It's a self-worth-building coping strategy, so the goal of being *our* best is converted into being *the* best, seen as *better than others*. When we're resorting to competitive behavior, we're often unaware that we're radiating defensiveness, jealousy, or envy.

Perhaps this situation sounds familiar: You head out for a fun night with friends. Over dinner, one person starts telling a story about their recent holiday. They get two sentences in, take a breath, and immediately another person jumps in with their own story, a "better" story about their "better" holiday. Maybe you've been the first person. Maybe you've been the second. Neither feels great, even if we can't pinpoint why. And that one moment can change the tone of the entire evening. Somehow, a fun night of bonding be-comes a little tense or emotionally fraught. Remember in chapter 2, when Eva's competitive behavior with her son ruined an otherwise lovely night with her family?

In our work, we get the most pushback on the idea that compet-itive is a below-the-line, ineffective behavior. Competition is good, we hear. Yes, it is: we believe the spirit of healthy competition is achieving. But unhealthy, pride-driven competitive behavior is not the same thing as above the line winning—working to be our best while achieving with purpose. Competitive behavior is not about winning with dignity, respect, honor, or integrity. It is more often about beating others and winning at all costs to make *ourselves* feel good and worthy. Think about some of the most inspiring sto-ries you've heard from the world of professional sports—stories of people or teams that have won with integrity or been gracious in defeat—versus those stories that have ruined careers and reputa-tions through cheating, drug-taking, or match fixing. The differ-ence between them is the essence of humility-driven achieving and pride-driven competitiveness.

Controlling behavior is a coping strategy that overrides our deeper positive intention to contribute to good results. Driven by pride, we become deeply attached to outcomes and what those outcomes will prove about us. *Controlling is the counterfeit of reliable.* It's about wanting things to be done well and "right," but not trusting that others can do it as well as you, so you micromanage everything.

Sitting between striving (described next) and competitive behaviors, controlling—when taken to an extreme—makes us "control freaks." When you're trying to control everything, you're actually *out* of control. We become focused on making sure everything goes according to *our* own plans, that tasks are completed *our* way and on *our* timelines. When we are triggered and pulled below the line into this behavior, we become obsessive about details. We may get results, but stress levels and antagonism are high. We exude dominance or arrogance, and an intense energy that triggers fear in others, even more so if we have "position power" either in a family or in the workplace. We also send the message that we don't trust people to make decisions, so we limit their growth as effective problem solvers.

As we've said, all these 16 behaviors are very common, so we can almost guarantee that you've had a controlling boss or colleague. It can certainly show up in parenting, too. On visiting her mother a few years ago, Mara was greeted at the door with, "It's cold! Why aren't you wearing a sweater?" Not exactly a warm welcome! Mara's mom's controlling behaviors could sabotage important moments. So you can see why Mara's reaction to Tamara's "You're just like your mother" comment was to clutch her heart and silently yell *Noooo!* Our coping strategies aren't a conscious choice, but essentially, Mara's mom was thinking and feeling: *If Mara's not protected against the cold, I'm a bad mother.* It was a small thing, but it was a template that threatened to drag Mara below the line. Instead, she took a breath, consciously shifted into compassion, and just laughed and hugged her mother. (Okay, maybe there was a small eye roll and some gritting of teeth, but she got to compassion after taking a *big* breath!)

Striving behavior emerges from a deep need to be right and to have an answer for everything. Pride overrides the good intention to achieve excellence and instead we become focused on avoiding all mistakes—and thus any chance of rejection. The resulting perfectionism can lead us to become workaholics, to be overly critical of ourselves and others, and to suffer a real loss of inner confidence if things aren't going well in some part of life. The template of "never good enough" can be rooted in striving.

How do you spell stress? S-T-R-I-V-E. Striving is below the line, achieving is above the line. When we're caught in a vortex of striving behavior, we radiate intensity, anxiety, tension, and a lack of joy while achieving tasks.

Morgan's mother, Nancy, leads a life that is defined by striving behaviors, or limited by them. Her sense of worth and confidence depend on making sure that everything is perfect—in this case, just for a family dinner. But her stress and anxiety strike fear into the hearts of those she loves as she becomes demanding and critical. She is preventing the very things she wants most: to avoid conflict, to be loved, and to be seen as a good mother and wife. She can't see that she is addicted to cleaning and perfecting her environment to calm the chaos of doubt inside. Unfortunately, her family struggles to be honest with her about how her behavior is making them feel. They have seen her defensive reaction of becoming even more critical whenever they try to give her feedback—all Nancy hears is "You're not good enough."

Striving is an important issue in our modern world, and many of us struggle with it. The so-called impostor syndrome is about striving—beating yourself up because you feel "never good enough" despite your accomplishments, and desperately fearing someone will find out you are a fraud. Overcoming striving often begins by recognizing the precursor: fear and our desire for approval.

The Four Self-Limiting Fear Behaviors

When the heart is operating out of fear, it manifests self-protecting behaviors that limit us like passivity, inferiority, and the need to please. Behaviors rooted in the fear quadrant are approval seeking, easily offended, dependent, and avoiding.

Approval-seeking behavior emerges from a good intention of wanting to be liked, to be accepted, and to get along with others. But fear overrides the love and creates a coping strategy of being overly nice, overly helpful, and overly agreeable. The desire for approval is innate for all of us. But people who struggle with a fear of rejection and sense of inner worth can develop a deep *need* for validation that shows up as people-pleasing. They use up emotional energy worrying what others think of them and run through mental scenarios practicing what they'll say in different situations so that it all comes out just right. They can radiate false flattery and false humility.

A couple invites some friends around to their new home. As the guests walk in the door, they immediately begin exclaiming, "Wow! This house is amazing!" even though they've only seen the entryway. As they walk through the living room, they release a barrage of overly effusive compliments, "That's awesome!" "Isn't that incredible!" As they move into the kitchen, one says to the other, "Darling, we need to get a fridge like that one!" One of the homeowners says they want to paint the room, and the guests enthusiastically agree. But the other owner disagrees. The visitors madly backpedal: "Oh, yes, we like it the way it is." But what do they *really* think?

We all slip into approval-seeking behaviors sometimes. But if we're living with this behavior as a regular coping strategy, it can lead to real problems, causing us to overcommit, say yes when we should say no, and struggle to establish priorities in our lives. It robs us of our authenticity (to ourselves and others), and can lead others to distrust us, as they sense they never know what we *really* think.

Easily offended behavior stems from the good intention of wanting to be encouraged, praised, and supported by others. But when we don't get the kind of positive response we desire, fear makes us perceive *feedback* as *negative criticism* and *correction* as *rejection*. We see it as a *personal* attack. We become overly sensitive, and so people often stop giving us any feedback at all, walking on eggshells around us. Without any proof otherwise, we think we're doing okay and our self-worth remains intact, but easily offended behavior is deceptive and can actually keep us from growing. People who score high on easily offended often radiate a prickly attitude and have unstable relationships because others feel they can't be honest, authentic, and relaxed around them. This is how Nancy's family felt around her.

Most of us have seen how this behavior can play out in the workplace. You're in a meeting where your thin-skinned colleague is presenting an idea or strategy. The tension in the room rises as everybody else is thinking about how difficult it will be to offer feedback. As the first brave soul speaks up, the presenter's expression becomes stony. By the end of the meeting, even though people have tried to be excessively diplomatic (i.e., not completely honest), the person has clammed up, become a bit sulky, and leaves without speaking to anybody. It can be exhausting for everyone.

Dependent behavior is driven by a fear of being excluded or rejected that overwhelms the good intention of wanting to serve others and do the right thing. It can seem easier and safer to let others take the lead rather than risk putting forth our own ideas. We don't have the courage or confidence to trust our opinions, our perceptions, or ourselves. We let others make decisions, follow their direction, and don't speak up. People who score high in this behavior can radiate indecisiveness or passivity. They don't trust themselves enough, so they keep checking themselves endlessly.

Do you have a friend who seems to let their spouse or partner make all the decisions? When it comes to where they're going on vacation, what color they're going to paint the living room, what movie they're going to see on Friday, they don't seem to have a say. You invite them out for a drink, only to hear "I'll have to check with . . ." You sigh and shake your head. And yet you've seen that same person be highly decisive and authentic about opinions and ideas at work. It's mind-boggling. Remember S+T=B—we're triggered into below the line behaviors in some situations and relationships and not in others, depending on what's happened to our hearts and how it has shaped our thinking—those templates! We can be dependent at home but not at work, with our parents but not our spouse, with a group of friends when we're not confident. A lot of times it's S+T=B—do we feel confident in an area? If not, we use dependent behavior as a coping strategy.

Avoiding behavior stems from a good intention of not offending or upsetting others, but fear of rejection keeps us from dealing with conflict or from taking any risks. As a coping strategy, avoiding behavior can keep us from holding ourselves accountable or being authentic with ourselves and others. Problems go unaddressed because we decide to "just leave it alone," to withdraw, to be superficial—we "brush it under the carpet" as the saying goes. We end up creating *artificial harmony* and never get to the heart of the matter.

Could you see in our example that Morgan's dad created artificial harmony to keep the peace by living in avoidance? Maybe that was the wise thing to do in that particular moment of a family dinner, but as a loving husband that kind of avoidance is not helping Nancy or the family dynamic. Have you ever had a serious challenge with a coworker who wasn't pulling their weight on a project, but instead of speaking up you just did more of the work yourself? Do you procrastinate on big projects because you feel overwhelmed by the responsibility? Have you ever "ghosted" on a friendship because there was a conflict that you knew would have to be addressed and you just couldn't deal with it? Avoiding behaviors are incredibly common in all aspects of our lives, as we work to feed our sense of worth.

We know intuitively that our fear-based and pride-based behaviors can cause us a mountain of stress and anxiety, and less stressful options are available. To grow away from our below the line and strengthen our above the line behaviors, though, it helps tremendously to understand where we are at right *now*.

How Are You Living Your Life Right Now? A Compass for Growth

We often feel the gap between where we want to be and where we sometimes find ourselves in our heart attitudes, our thinking, and our behavior.

We believe the Heartstyles Indicator helps shine a light on the reality of how we *currently* think and behave. It helps us see how effective we are right now. And just as important, it helps us define how we would *like* to be living.

We always emphasize that the results are not a tool to judge but instead create a personalized *compass for life*. The results can change as we develop our character, and so they should encourage us to begin that evolution. As you complete the online questionnaire (if you haven't already), you will respond to seventy-five statements twice. The first time you will respond based on how you would like to behave, what you value, or what you *aspire* to—this creates your "Benchmark." The second time you respond based on how you think you're *currently* behaving. This creates your "Self Score." When you are done, you can download a personal development guide with your results and an explanation of how to read them. They will be captured in graphs like the ones below. When you look at the indicators side by side, any difference between how

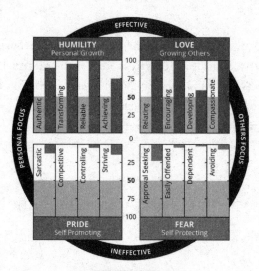

BENCHMARK

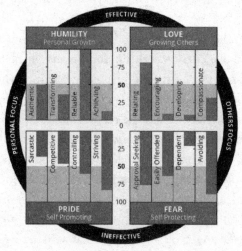

SELF SCORE

you would *like* to be living and how you're *actually* living can be very evident.

Let's return to Morgan, from the beginning of the chapter. She took a 360-degree version of the Indicator, which meant that her report also included averaged responses from three of her peers who also completed the same seventy-five questions on her. So, Morgan had a Benchmark, Self Score, and Others Score. What do you think her results showed? On the previous page are her Benchmark (how she'd like to be), and her Self Score (how she thinks she currently is), while below is the Others Score of her averaged peers' responses on their perception of her current behavior:

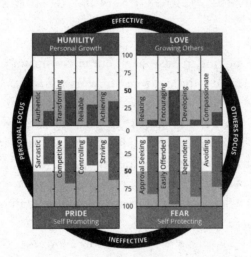

OTHERS SCORE

Morgan scored particularly high in fear-based behaviors. Nancy's pride-driven behaviors provoked self-protection in Morgan—she often felt criticized or not good enough. With the insights she learned from her results, Morgan could see the false security she was building through her coping strategies and began to choose to change her patterns and lessen her fear-based behaviors. With humility, she admitted she was living out of fear, not just at home but at work, especially when she was around other controlling and striving people. Now she understood her templates and how other striving and controlling people triggered her—S+T=B! She also saw how her own need to self-protect was impacting others around her at work—she wasn't seen as warm and interested in others—and that really floored her. She started to focus on strengthening her personal growth behaviors, working on becoming more authentic and achieving, and this helped her become more of herself with others, who felt the positive impact of that. It made a dramatic difference in her family and work dynamics and her life as a whole.

Living Inside Out, Not Outside In

Our below the line behaviors are based on an *outside-in* approach to the world, looking to others and our environment to approve of us, validate us, and build our sense of worth. Because our need for approval is so innate, we are often triggered into approval-seeking behaviors. When we need more than what those behaviors can deliver because we have voids in our hearts that we're trying to fill, we often resort to pride-based coping strategies. When triggered, we attempt to prove our worth by controlling our world (striving and controlling) or demonstrating that we're better than others (competitive and sarcastic).

If we haven't gotten the approval we're looking for in life and carry wounds of rejection in our hearts, we can be easily triggered into the fear-based behaviors to protect ourselves. Being offended is often our first line of defense, but being dependent and avoiding are also coping strategies that protect our self-worth and feed our need for approval. In our modern world, these behaviors can deliver the approval we're searching for in many forms: money, promotions, trophies, and compliments.

The outside-in life is a set of coping mechanisms for proving and protecting, not strategies for thriving and growing. Those coping mechanisms never fill the void or heal the wound, though, and so we keep repeating the behaviors. The sad thing is, living like this won't build or strengthen our *character*. In the workplace, the irony is the more we try to protect our job by using the self-protecting behaviors, the more likely we are to be retrenched in a restructure. The more we try to promote ourselves by using self-promoting behaviors, the more we can be seen as a "culture-buster" and not achieve the promotions we think we deserve.

The *inside-out* approach to life* is when we have a strong sense of our inner worth and purpose. We build our character from the inside, to know who we are (authentic and transforming) and know where we're going (reliable and achieving). We then have the desire to build that same sense of worth and purpose in others, to move forward alongside them, which shows up in the growth-driven love behaviors (relating, encouraging, developing, and compassionate). Above the line behaviors bring the same rewards in life in terms of the usual measures of success, but the fulfillment and energy we get from above the line behaviors are a stark contrast to the exhaustion, stress, and anxiety that often come with below the line behaviors.

We can choose to shift our hearts, our thinking, and our behavior above the line, from the inside out. That is the choice we face moment by moment in the cultures we live and work in that tug on us day by day to live below the line. The more often we make the choice, the stronger we get and the easier it becomes—until it is our default way of operating . . . most of the time!

In the next chapter, we will piece together the last four chapters into case studies, so you can see what all this looks like in day-to-day life, and how it all plays out moment to moment.

♥

CHAPTER 5

Spotting Behavior Patterns and Connecting the Dots

Two months into the rollout of the audio manufacturer's new range of headphones, and it wasn't going well. Missed revenue targets and underperforming new headphones were just the start. To top it off, customer complaints had been streaming in because of a technical fault with the volume control.

Theresa leaned over the small conference table, her hands curling into fists as she eyeballed her colleague. "We wouldn't be in this situation if we had gotten off to a stronger start."

"What do you mean?" Darren, head of product development, couldn't believe Theresa was trying to lay this at *his* feet. His team was one of the best performing in the company. His face felt hot and he was barely controlling his volume. "We hit our deadline *and* delivered a great product. We gave you what you needed to finish the marketing rollout. If your people couldn't deliver . . ."

"My people did a *great* job with what they were given," she responded. "But *your* people weren't entirely helpful when my team went hunting for the information they really needed. And *now* we have to deal with this technical problem, which certainly isn't *our* fault."

"Ooooh! You two sound like our angry customers!" Bettina, head of sales, chimed in with a voice dripping with sarcasm, but softened the comment with a smile. She pulled up sales numbers broken down by region and outlet. "I've asked you all for updates so that I can revise sales targets. I need that information."

Theresa heaved a frustrated sigh. "Bettina, I don't know what else you want. I've sent you everything we have."

"I need a deeper dive into the marketing data if I'm going to turn things around." Bettina had to have *all* the details so she could solve the problem. Tim, the CEO, had asked the team to develop immediate solutions for the manufacturing issue and new marketing and sales plans for the next quarter, and she wasn't about to drop the ball. *It won't be me that lets him down—no way will I let that happen,* she thought.

Darren broke in. "I think we need to move *now*. We can't waste more time gathering information. I've developed some ideas to deal with the manufacturing issues."

"We need to discuss everyone's viewpoints, Darren," Bettina replied. "Whatever plan we decide on affects all of us, not just your team."

"Have you even discussed your ideas with Mark?" Theresa asked. "He's the one who has to make any plan work."

Mark, responsible for operations and manufacturing relationships, had been silent for most of the meeting, leaning back slightly from the table, gauging the rise in tension. He could see a sound solution in front of them, but he held his tongue. *It might not work,* he thought. "I haven't seen them yet, no, but I'm sure Darren was just vetting them with his team before getting my input."

Theresa made a dismissive noise. "You shouldn't be waiting for an invitation to be part of that discussion."

Round and round they went. At the end of an hour, they were all frustrated and only had a rough framework for a plan. They all felt the pressure to move faster and head off more poor results. The blame game continued, and they just couldn't seem to get unstuck. The good news was they were all passionate about the brand, product, and wanting the same thing—to fix the problem and get sales up.

After reading the last four chapters and completing your own Heart-styles Indicator, you probably recognize some of the coping strategies these four people are struggling with. Theresa, Darren, and Bettina all score fairly high in some of the self-promoting behaviors. Theresa and Darren's competitive is causing them to throw down blame because neither wants the poor results to cast a shadow on *their* performance—their sense of worth. All three are high enough in controlling that their ability to work together to solve the problem is compromised. And with Bettina's sarcastic and Theresa's striving, tensions during discussions get high. Meanwhile, Mark's higher dependent and avoiding behaviors keep him from taking a risk and speaking up with ideas that could help the team and the company. At that meeting, they are all just reacting to each other's below the line behavior.

We work with many teams who struggle with these types of dynamics. Commonly, as soon as a situation of performance, crisis, or possible financial loss happens, it triggers fear, thus below the line behavior. Striving, competitive, and controlling behaviors are especially common for leaders, because those behaviors can, and do, deliver results. But the personal toll, and the toll on culture and team performance, can be high. Taken to an extreme, these behaviors become arrogance—which is when objectivity, innovation, and nimbleness disappear. People become defensive. Living below the line in this way costs energy and time; it can lead to sloppy decision making, counterproductive game playing, even corporate backstabbing. Translate those same behaviors into families, or other environments, and the effects can be just as harmful. Basically, where we have people, we will have some degree of coping strategies that cause mayhem in relationships—but we *can* rise up above the line!

Usually, the first thing we hear about the challenges a team is facing is blame: their difficulties are caused by a stressful project, a crisis, or "clashing personalities." But individual behaviors guide how we respond to crises and stress and conflict. Below the line coping strategies can often trigger the same in others, setting off a downward spiral that keeps us from moving forward with positive purpose. Maybe you've seen that in your own family dynamics or even social situations. For so many of us, the words "he started it" bring back pretty vivid childhood memories!

Yet for any person or team, those below the line behaviors don't paint the whole picture. Mark, for example, was known for being very trustworthy, something his colleagues valued in him, as well as being an all-around likable guy whom people enjoyed working with. He scored high in reliable and relating, and strongly in authentic and compassionate behaviors. Bettina was also strong in the love quadrant and was well-liked. Theresa scored very high in humility-based behaviors, especially authentic, transforming, and achieving. She was respected for her focus on excellence. Darren produced impressive results for the company; his colleagues appreciated what he contributed and that he got things done.

The question, though, is what happens when these people get trapped and triggered into below the line behaviors? Why do they do what they do, how do those behaviors show up in different areas of their lives, and how can they gather insights that help them shift above the line?

Here we're going to show you how this looks in the real world and the power of shifting away from below the line behaviors. In this chapter we're going to explore how the pieces fit together, encapsulating everything we've done together so far. The four principles; S+T=B; triggers, templates, and truths; and the voids, wounds, vows, and gold of the heart create patterns of behavior that shape our lives. You have an opportunity to practice some grace for yourself and others as you are learning about what lies behind why we do the things we do!

As we look inward to explore our own behavior patterns, it can be easy to overlook or misjudge how our behavior affects other

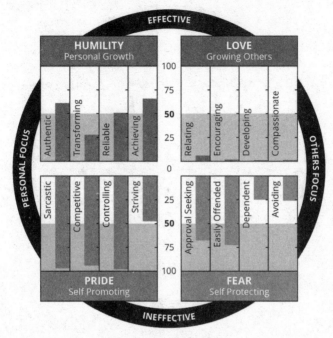

DARREN

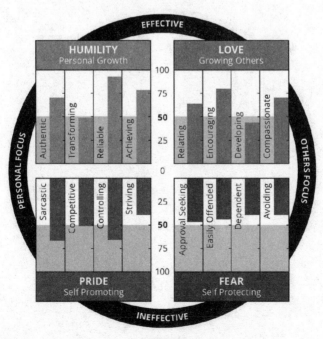

BETTINA

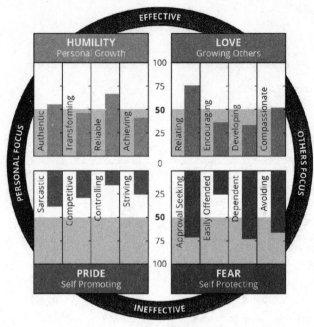

MARK

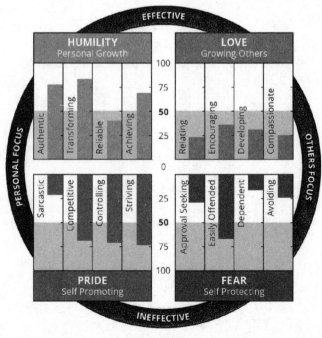

THERESA

people. Just as we can be triggered by others, we can trigger *them* into below the line coping strategies—or we can provide them with the opportunity to rise into above the line character growth. Understanding our role and finding compassion for the people in our lives with their own triggers and templates and personal journeys of growth is key to our own growth.

The people we describe in this chapter, and their journeys, are all based on real individuals we have worked with, and some of their identifying details have been changed for their privacy. We hope their stories will unlock insights for you and help you spot opportunities to shift your behavior and positively influence those around you.

As you read this chapter, have your own completed Heartstyles Indicator beside you, so you can read what's happening in the story and relate any relevant behaviors or history back to elements in your own life. Reflect on the interplay between your own life and how your behavior may trigger the people you work, live, and socialize with—and vice versa!

Connecting Your Behavior to Your Head and Heart

Not long after their product launch problems, Theresa, Darren, Bettina, and Mark were part of a Heartstyles workshop and coaching program. Tim, the CEO, could see the team challenges among his leaders as well as the rest of the company and had decided to do something about it. So, at a group workshop, the team members received their Heartstyles results. Their Self Score results showed how they rated themselves, as well as the averaged results of their Others Score (they each chose nine respondents to complete the Indicator on them). For some of the team, their results were not surprising. For others, it was a different story.

The Only Way to Get Ahead Is to Prove I'm Better Than Others

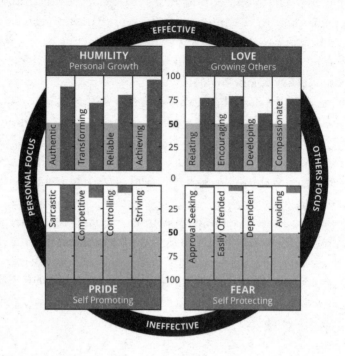

When Darren got home that evening, he was on edge. His fiancée, Shannon, asked what was wrong, but got a blunt response. She prodded anyway, despite knowing she might be bringing on an argument.

"I just don't understand where my team is coming from," Darren blurted out. "Look at this. Do they really think I'm not at all encouraging . . . or *compassionate*?" Waving his PDG in the air, he threw it on the sofa as his voice took on a hard edge.

Hearing that tone she knew so well, Shannon hesitated, trying to come up with a diplomatic answer. They had been together for two years, and as they moved closer to marriage, she was having doubts. She loved Darren, but he just wasn't there for her when she had a tough day. He had a lot of attributes she appreciated, but she knew not to expect him to take care of her when she was sick or surprise her with her favorite flowers. Could she live with that— forever? This wasn't the time for that conversation, but she wanted to be honest with him about *this* feedback he had received, because she understood where it was coming from.

"What specifically do you do to show encouragement?"

"What do you mean?"

"Do you tell people when they've done a good job? Do you praise good ideas? Do you show that you're there to support them?"

"You mean *pamper* them?"

"Well, I guess that's an answer."

Darren felt his face get warm again. It brought up memories from the disastrous team meeting and a discussion during the workshop. *I'm being triggered*, he thought. *What Shannon is saying is actually triggering me. Why?*

He had become very uncomfortable when they began to talk about the voids and the wounds of the heart in the workshop. He could see the value of S+T=B, but work wasn't the place to complain about your last bad boss or share sad stories from your childhood.

Very few people in the company knew that Darren's father had died when he was twelve. He had sat in the hospital waiting room with his mother and his three younger siblings, waiting for his dad to come out from emergency heart surgery. Eventually the surgeon came to tell them his father had died on the operating table. Darren, understandably, felt helpless.

Despite the responsibility of helping his mom raise his siblings, Darren had done well in school, needing to prove that he could succeed despite the weight on his shoulders. The void of fatherly love and the diminished attention from his struggling mother created competitive and controlling behaviors that had stayed with him his whole life. His father's death and his own experiences thereafter created inner vows of "It's up to me; I'm on my own," "Life's short, so do well fast," and "Don't let your feelings take over, it's too painful." These inner vows formed in adolescence created the foundation for templates that helped Darren interpret most situations as an opportunity to prove his self-worth, in one way or another, and to build a wall around his heart so that emotion wouldn't overwhelm him. Sarcasm was an easy way to build that wall in social situations, where he felt unable to connect or where he felt people were getting too close to his vulnerability.

Darren's heart attitude, thinking, and behaviors helped him do well in his first professional job where the company culture emphasized performance and results as the only measures that mattered. There was no real people development. That suited Darren fine—he didn't have to face his own personal development and he didn't have to open himself up to caring for others by helping *them* develop. But as a leader in his current organization with a very different culture, Darren was struggling. And his Heartstyles results put him in a position to examine why.

While Darren recognized that his early life had had an impact on him, he wasn't ready to acknowledge the connection between his heart, his head, and his behavior just yet.

If he wanted to salvage his career in that company, and possibly his relationship with Shannon, he would have to—soon.

If I Lose Control of This, Something Will Go Wrong

Darren didn't know it, but Bettina had grown up without a father as well. She had never known her father at all, in fact. Her mother gave birth to Bettina when she was just sixteen, and her father had moved away soon after, refusing to have any part in Bettina's life. In her home country, there were few laws requiring child support. Bettina's mother worked three jobs to make ends meet.

Bettina's grandparents were ever-present in her life, and between her mother and grandparents, she always felt loved and cared for. But it didn't change the shame she felt of having a father who had abandoned her. To explain his absence, she would tell people that her father worked in a neighboring province and could only make it home on the occasional weekend. All through her childhood, the need to keep up this fiction made it hard for her to let anyone fully into her life. Focusing on being accepted, despite her family

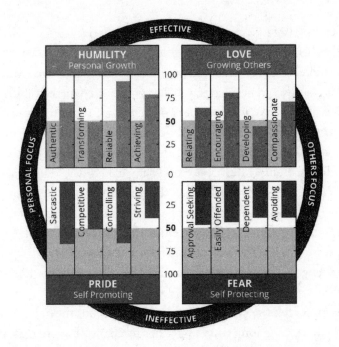

situation, she would use sassy and sarcastic humor to stand out and build friendships with her peers.

Her inner desire to honor the sacrifices of her mother and grandparents led Bettina to do well in school. The more they praised her for being smart, the more she used information as a way to be seen as worthy. The more information she held, the more in control she felt.

Bettina had met her husband in college. They had been married for ten years, had a daughter and a son, and had built a close, loving family. At home, Bettina felt loved and accepted for herself, so her pride-driven behaviors of sarcastic, competitive, and controlling were rarely triggered and were balanced out by her love-driven strengths of relating, encouraging, and compassionate and her high humility-driven behaviors. At work, on the other hand, S+T=B applied: Bettina's need to do well and prove herself meant that she was regularly triggered by colleagues who had similar below the line coping strategies, like Darren and Theresa. She had been head of sales for three years and a sales leader for six years before that, and had been instrumental in building many of the processes and tools the team used. Her knowledge was deep and wide, and she liked it that way. Knowledge was her go-to self-worth stabilizer.

It wasn't so easy for members of her team, though. New members, with experience from other companies and ideas for how to improve or innovate how the sales team operated, felt that she shut them down. Behind Bettina's back, they called her "The Queen of No." They liked her, and they felt that in many ways she was a good leader. If you were struggling or had a personal issue, she was there for you. She knew everybody well, understood their strengths and weaknesses, and pulled them together as a team. She didn't like to be challenged, though, and that made them feel they couldn't fully contribute.

Her peers felt the same. They liked Bettina, but they also knew it could take a long, long time to get her to change how her team operated, to cooperate with other teams, to share details on sales performance, and so on. They just wanted her to be more collaborative.

Keep Safe by Doing What I'm Told

For days after the meeting, Mark kept pondering that idea he had for solving the manufacturing issue. Darren had shared *his* idea, at a high level, and Mark knew it could work, but it wasn't perfect. When Mark asked questions, Darren blew them off or made snide remarks. And given his dependent and avoiding coping strategies, that made Mark retreat.

When we struggle with fear-based behaviors, we're often triggered by sarcastic and controlling behaviors in others. The more they advance, the more we retreat, fearing that they'll lash out and criticize or reject us.

It didn't help that Mark was new to his senior role. He didn't want to mess up in the first six months, so he kept second-guessing his decisions. It was exhausting. He remembered his father's wisdom

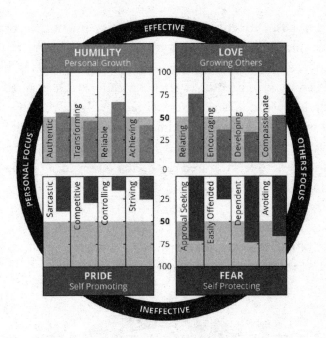

almost daily: "Keep your head down and work hard." (His father had worked in middle management for thirty years with the same company.) Two previous jobs at other companies that had aggressive cultures cemented this idea and contributed to Mark's inner vow, "Don't take risks or stand out; it's too dangerous." Mark had been working hard his entire career, and that showed in his reliable scores. He was a known quantity, and it had helped him get promoted to lead a well-established team that thrived on process and systems. But as a leader, he had to take initiative, he had to advocate for his team, and he had to commit to ideas that would help the company make positive progress. So far, that wasn't happening.

When Mark reviewed his Indicator, two days after the unproductive meeting, the high dependent and avoiding scores stood out. He could connect these results to his behavior in that meeting and recent feedback he had received from Tim, the CEO. He had always been afraid to ask for help, especially in new positions, but he didn't know how to break the cycle. And he could see that the lower scores in encouraging and developing were problematic if he wanted to be an effective leader for his team and in this company. "Maybe this feedback has come at the right time for me to finally let go of some of those inner vows around playing it safe," he realized. He wanted to become an effective leader and he wanted to get the help he needed. That desire was getting bigger than his fear of asking for help—maybe it *was* time to do something about it. Tim had been a great coach with his feedback—he had created a safe place for Mark to feel inspired to develop rather than feeling afraid of letting everyone down.

If I Don't Push to Prove, Perform, and Perfect, I'll Lose Everything

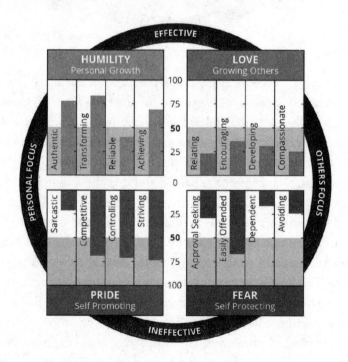

Theresa was an avid learner, and she saw any opportunity to grow and improve as a gift. It wasn't much of a surprise for her, then, when she received her results. She hadn't thought of her behavior in those terms, exactly, but she knew from feedback she had received in the past that her peers found her to be combative at times, and a bit scary when she got that way.

Thinking back over her life, she knew they were right. Theresa's was a real-life rags-to-riches story. She had learned to fight young, as she had grown up in poverty. Her parents wanted better for her, but they were both uneducated and raised in poverty, too. Her family's situation isolated her, and through her first few years at school she was often made fun of for her shabby appearance. She was gifted with a clever mind, and she devoted herself to doing well in school, determined to escape the poverty she was living in. When she wasn't looking after her siblings, she was studying. She earned a scholarship to attend high school and then university. She was the first in her family to finish high school, let alone university.

Theresa could have told you that her striving behaviors came from a deep desire—an inner vow—to never be poor again. It drove her to push herself to perform. Her experiences with bullying created two templates: that you have to fight for everything in this world and fight others first before they fight you; and that the downtrodden or the underdog needs to be protected, because she often wasn't. So despite her intensity, her team loved her because they knew she always had their backs. Her colleagues, though, felt she fought for her team *too* much, in a way that created conflict between departments.

It bothered her to see the low scores in the love quadrant. Her relationship with her husband was strong, but she didn't have the intimate friendships some people seemed to build. She had one or two close friends, but she knew she didn't share much with them about her deeper feelings, her challenges. *Why should I?* she thought. *They could use it against me sometime.* One friend had been in the hospital recently, and she hadn't had time to visit. Okay, if she was honest with herself, she didn't *make* time to visit, as she put her work deadlines as first priority—what the business was going through *was* important. Wasn't it?

Look Inside—Learn *Your* Story

We all have a story that will have shaped our thinking, our behaviors, and our context of how we see the world—*our* truth. We, too, like Theresa, Darren, Mark, and Bettina, are an AND—our behavior in any moment will usually depend on the context: S+T=B. The question is, though, can we begin to operate more above the line, more of the time, when we get triggered into potential below the line coping strategies? If we can learn what our triggers, templates, truths, and inner vows are, we have the opportunity to be able to remain above the line across most of our contexts!

It's reassuring to think that *we're all in this together* because even though we might be different people with different backgrounds, we all have similarities. Two people in a room may both have competitive coping strategy behavior, which might have come from very different origins, but the behavior is similar even if the templates are not. Knowing that this is so makes having compassion for our colleagues, friends, and family—even random strangers in the traffic—so much easier.

Seeing the below the line behavior driving the dynamics within his leadership team and eroding their effectiveness, CEO Tim was highly motivated to turn around the culture for the better. A talented leader with a background in finance, Tim was a product of above the line development. He had developed from a brash and competitive young executive into an authentic and compassionate heart-led leader who got results while also being known for his inspirational coaching and development of those around him.

Tim had been very influenced by seeing the cumulative Heartstyles data (more than 100,000 respondents). He particularly focused on the relationship between the "effectiveness at work" question with the low to high scores in above the line behaviors.

Fear and pride do not make for a healthy work culture. Tim knew that, and at the time Darren, Theresa, Mark, and Bettina met, they, too, were just beginning to get a sense that shifting toward

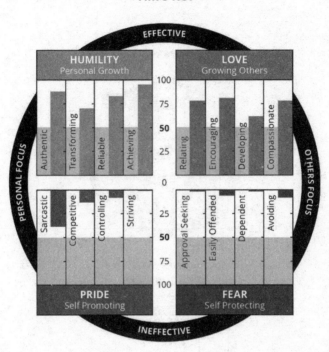

Tim's HSI

humility and love might make the difference. The only thing was, given their history, how would they do that?

No More Game Playing—Let's Connect

Theresa had a heartfelt epiphany at dinner with her husband speaking about her feedback—particularly the very low love quadrant results and how she realized her templates were getting in her way. The day following their team meeting, Theresa made a decision to be vulnerable and try a different approach with the others. She walked into Darren's office. "We should talk," she said.

Current level of effectiveness at work

Respondents who scored participants **very low** to **moderate** (n=21117 : 17%)

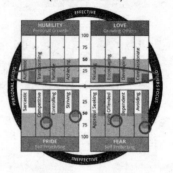

Respondents who scored participants **high** (n=54332 : 45%)

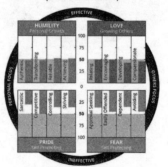

Respondents who scored participants **very high** (n=45474 : 38%)

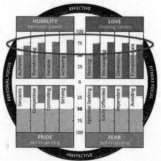

"I've been thinking it's in all of our best interests to get this rollout back on track. What we discussed at the workshop about working together—*really* working together and not letting ego-pride get in the way—what if we started practicing that?"

Darren eyed her with a mixture of anticipation and cynicism. "Sure, I'm glad to hear you're finally ready to play ball on this."

"You know, Darren, I don't want to play games anymore. How about we wipe the slate clean, get Bettina and Mark in here, too, and work through the ideas we all have to fix the situation."

Darren raised his eyebrows slightly at Theresa's resolute tone but felt himself nodding in agreement. "Yeah," he sighed. "I know what you mean. I'm tired of butting heads. I know we're all after the same thing." He picked up the phone and called Mark and Bettina, asking them to join their impromptu meeting.

When all four of them were in the room, Theresa again took the lead. "Guys, Darren and I have realized that we all want the same thing—for our company to be successful. I'm sure we all have some valid strategies to iron out the problems with the headphone labeling and salvage what we can of the launch. I propose that we practice what we learned in the workshop about being above the line, and we go around the table and lay out the ideas we each have."

"I like that," Bettina agreed. "One proviso: everyone has to hear each other out. No interrupting, no digs, no putdowns. Let's keep it above the line."

The others nodded.

"Well, in that case I'd like to go first," Bettina continued. "We've been working on strengthening our regional sales strategy, but to be honest we're struggling to really make an impact on the numbers. *I'm* struggling." She took a deep breath and watched her colleagues closely. She had never admitted vulnerability to such a degree in the work environment before, and she felt a little jumpy, anticipating some kind of backlash.

"I think I can see a way to deal with that," Mark said thoughtfully. The others looked across at him, a little taken aback to hear him contributing so early in the discussions. They were used to him holding back until the closing minutes of any meeting before speaking up. "I've been thinking that we could get the factory to expedite shipping the relabeled headphones into regional centers, while the marketing team focuses on the Chicago audio conference as a possible relaunch opportunity."

Darren leaned back in his chair. "I would have thought you already had the factory pulling out all the stops . . ." He caught Theresa's eye and paused. "Sorry, Mark, that was out of line. If you think you can bring some influence to bear, that makes good sense."

Mark, surprised but cautiously pleased at Darren's apology, went on, "Thanks Darren. Yes, I believe this plan can work. I propose we go for it!"

"Great—that buys us some time, and it'll give us the chance to really do something special in Chicago for our key influencers." Bettina smiled.

"Can you believe that we've been sitting in this room for less than thirty minutes, and already we've agreed on some viable ways forward!" Theresa marveled.

"Don't get too comfortable," said Darren. "We haven't actually *done* anything yet." *Oops,* he thought, *there's my sarcasm again. Let me try again.* "Sorry, guys. I know that was sarcastic. I was just trying to lighten the room. And," he said looking down, "I appreciate you all for your efforts."

"Thanks, Darren," Theresa replied. "Maybe if we can keep being honest with ourselves like you just did, and with each other, like we have today, we can start to change the way we work together. Our teams will take our lead. Maybe we can put a stop to all the dysfunctional interdepartmental game playing. Maybe we need to fess up to being the leaders of it in trying to protect our turf and our teams."

Through a series of meetings, they came to the decision that this group—not their own functional teams—was their number one team. As hard as that was, it was probably one of the main factors allowing them to develop into a cohesive team. They created and agreed to some behaviors that were nonnegotiable between them. They agreed to call each other out when someone went below the line into one of their coping strategies, and they agreed on how each person wanted that done. They committed to having each other's backs, and were called to practice that quite a few times! Their fledgling trust of each other took root, and even when they had some ups and downs, they kept true to their commitments. Eventually, they could smile at the sarcasm moments instead of taking offense, laugh at the sitting-on-the-fence moments, and ask questions to draw out the solutions that were there. They could practice S+T=B if tension started during meetings about tough topics. They had a poster of the Indicator in their meeting room to remind them of the styles, and four foam balls in the four Indicators quadrant colors sitting in the middle of the meeting room table, to use to give someone a hint if they went below the line or as silent praise when above the line. Okay, maybe there were a few lighthearted throws across the table at times—but that showed how far they'd come. As Darren told us, "Yeah, once upon a time, I would have been knocked unconscious if we had those balls before we knew all this stuff! I was a real pain!"

It took courage for the team to be authentic with themselves and each other—asking why *do* I do the things I do? Do I want to keep living like this—what's the cost? If some parts need to change, then what and how? It took trust and vulnerability, and the openness to see they were all in it together, and that as humans, they all had triggers and templates and vows that were stopping them from being their best selves. Above the line spirals above the line. Deep within, they knew they were resonating with their authentic, best selves. Now they're less stressed, they're saving time, and their decision making is more effective because they've all learned to interact in an above the line way (most of the time!). They have developed a trust for each other and they each know they have each other's backs. As a result, the culture that Tim envisioned is now in place. Departments aren't fighting and competing, and innovation and collaboration have increased, and it's always a work in progress!

Look back at that first meeting, how tempers were running high right from the get-go. Bettina was set on gathering all the information so she could feel in control; that made Theresa feel under attack, so she went into fight mode; Mark was in retreat from Darren's sarcasm; while Darren was striving to get results at any cost and ready to beat down anyone who got in his way. The group still has its ups and downs, of course—they're human! But now they can ask one of Mara's favorite questions: "Am I reacting to a current situation based on a past experience (template)?" By being aware of what is triggering them, where their templates have been formed, and the gold that they have to offer, each of the four is able to make a conscious shift to thinking and behaving above the line more often.

And Darren? He knew that the changes he was making at work were great, but he needed to make changes at home even more so. He asked Shannon to tell him where his sarcasm was costing him her trust. In Shannon's tearful reply, Darren finally heard all the feedback he'd been given many times over, from friends and at work. By the time they married, Darren had become a man his dad would have been proud of: a high achiever, yes, but one full of integrity, who loved developing his team to greater heights than he knew *he* was capable of, who apologized and calmed down when he caught himself getting pushy in meetings. And he has become a loving partner and dad. Shannon gets flowers from him a lot—with little notes. Only a couple of words—but they are heartfelt.

You've been looking at your own Heartstyles Indicator results, reading about Mark, Theresa, Darren, and Bettina and the changes they have now made in their thinking and behavior. You may be feeling fired up to make those kinds of changes in your own life. But how? Where do you start? That's exactly where we are heading in chapter 6.

PART II

Shifting Above the Line

CHAPTER 6

Trade Up! How to Resist Being Pulled Below the Line

Welcome to part 2 of *Above the Line*. By now you're probably saying to yourself "Okay, I get it—but *how* do I live and lead above the line?" The next five chapters will equip you with the methods that you can apply in all aspects of your life. They are a combination of Heart + Smart: skills that strengthen your character and, in time, can become a natural part of you. Please remember we are not asking you to change your personality; we are offering you the opportunity to develop your effectiveness and grow in wisdom.

There's truth in the old saying of "take the best and leave the rest." Some things will particularly resonate with you as you go through part 2, so grab hold of these and use them. In another six to nine months, skim through part 2 again: perhaps something else will resonate with you then to use in your development.

Character development is just like going to the gym. The key is consistency and coaching. Physical strength isn't built in a day, and neither is character. And in both cases, it's helpful to have someone by your side as you apply the methods in a way that will really make a difference. Consider these chapters to be your personal trainer in the my-best-self gym. If you apply the learning over time, you will see massive development, growth, and good things happening in your life personally and professionally. Be consistent in what you're putting into practice and have a mentor, a friend, a colleague, or your partner to be your "truth-teller" providing you with feedback.

Now that you're comfortable with the terms we are using, from here on let's go with ATL and BTL for above the line and below the line (so much easier!).

The secret of trading up from BTL to ATL lies in three simple steps:

1. Identify—what's happening for me

2. SBTB: *Stop, Breathe, Think, Behave*—what to do in the moment

3. Plan—what to do in the future

Richard pulled into the driveway and felt his hands clench the steering wheel. "What the . . . [expletive deleted]!" *Why is it so hard to remember to close the garage door?* he ranted internally. *It isn't rocket science; it's one button!* He hit the brakes, got out of the car, slammed the door, and stormed into the house.

Susan couldn't believe her boss was supporting the new plan from the CEO. It wasn't going to work, she was sure, and was more likely going to backfire, hurting their sales and productivity. If they had asked anybody on her team, they would have learned what the problems would be—before they stormed ahead. Well, she wasn't going to help them get out of this one, and her boss could find somebody else to sell it to the team. She would tell him just that at their next meeting.

"Mara, where are the car keys? You had them last."

"No, darling, I didn't."

"Yes. You *did.*"

"No I *didn't.*"

You're making us late for the show!

Can you commiserate? This week, have you thought about laying on the horn after being cut off in traffic, or did your blood pressure rise when a colleague criticized your ideas in a meeting? Was your frustration at an all-time high when your car rental was not ready and the hotel booking was wrong all on the same day, or did you want to shout at your kids because they weren't getting ready fast enough? Or maybe there was something else pulling you BTL to a habitual coping strategy? In each of the scenes above, that's what is about to happen. It's a very normal *reaction* but not the most effective *response.*

Every day we're faced with situations that could trigger us to go BTL. Some of those situations are legitimately triggering—true wrongs or injustices in our world or life. Some may seem that way, but are actually perceived offenses based on "my truth" and not "the truth." And some are just day-to-day situations we all have to deal with—like being caught in traffic or running late!

You can tell the greatness of a man by what makes him angry.

—ABRAHAM LINCOLN

Every day, every hour, sometimes moment to moment, we each have the opportunity and the choice to trade up—to shift our behavior away from one of our habitual BTL styles to an ATL style. Richard, Susan, and the two of us certainly did. We just had to *recognize* the signs that we were in danger of resorting to ineffective behavior. Luckily, that happened.

Richard stopped as he got just inside the door and took a deep breath. He thought about why he was so angry about something fairly inconsequential. He was tired and knew that a missed deadline by a member of his team at work that day had made him frustrated and drained for the rest of the day; he had used up all his energy to stay ATL at the office. As he stepped into the house, he took an even deeper breath and made a choice to bring his loving self, not his carried-over coping strategy, to his family.

Before Susan spoke too soon with her boss, she thought about identifying what behavior she was about to launch into, for him to see the flaws in the new plan. Then she had the wisdom to stop and recognize that maybe it wasn't her truest intention. What she *really* wanted, she discovered, was for the team to continue to be successful, and she was sure that was what her boss wanted, too. What they needed to talk about was how to make that happen. Besides, if she went into the conversation attached to showing him how *wrong* he was—and how *right* she was—he wasn't going to listen . . . and it would no doubt end badly!

Mara and Stephen stopped the blaming and accusing over who had the keys last, took a breather for a moment to both calm down, and then hunted for the keys—together (sounds a bit too-good-to-be-true, but we did—*that* time!).

What surprises many people as they start to consistently trade up is the shift they see in others. "My boss seems to respect me more." "My spouse is more loving." "My kids are listening better." As we've said, ATL breeds ATL. Showing others the humility and love in our hearts encourages them to do the same. But *someone* has to *start!* Sometimes we just need to "put on our big boy/girl pants" and be the one to go ATL. Consider that if you ever find yourself thinking, *That will never work with my boss/ partner/kids/parents.* And even if others don't change, at least you can—and you'll come out of the situation more positive and feeling stronger because of it.

It's incredibly empowering when we trust ourselves to not be pulled BTL by other people's behavior or situations. Many years ago, a participant said at the end of a program, "I will no longer be ruled by old triggers and templates"—what a great positive inner vow! When we operate from calm (calm command, calm connection, calm compassion), we feel more confident that we can handle tough situations and stay true to ourselves. We learn to focus on and speak to what is most important and let the rest fall away. And we build the relationships we want as our compassion for, and understanding of, others grows.

Now let's walk through those three steps for trading up to ATL behavior:

1. Identify—what's happening for me

2. SBTB: Stop, Breathe, Think, Behave—what to do in the moment

3. Plan—what to do in the future

1. Identify—What's Happening for Me

Identify—what's happening for me is the first step toward trading up to ATL behavior.

Looking in the mirror of our own behavior isn't easy—we don't always see ourselves so clearly—so here are some tips to identify what is going on for you. Ask yourself these questions or go to your PDG (Personal Development Guide):

1. What BTL strategies do I use under pressure?

2. Remembering S+T=B, what situations need to happen for me to be triggered to go BTL?

3. What template is in play? What voids, wounds, or inner vows have been ruling my emotions and helped build that template?

4. What is *my* truth versus *the* truth here?

5. Am I tired or carrying negative tension from another situation?

6. What is my ego attached to that I need to let go of?

7. Do I have feelings of rejection, not being good enough, the need to prove and perform or to be right all the time when S+T=B?

8. What are the positive intentions of others? What is happening for them?

9. What is our shared meaning and truth?

10. Where can I extend grace to myself and others?

Feel It—Your Body Doesn't Lie

As we wrote in chapter 2, encoded in the templates that are stored in our brain are the memories associated with past experiences, but also the *physiological reactions* that happened at the time. Those reactions stem from the fight-or-flight mechanism in the brain. It causes the release of epinephrine (or adrenaline) to boost our energy and strength and then, if the threat hasn't disappeared yet, the release of cortisol (recently labeled "Public Enemy No. 1" in an article headline we saw!) to keep us in a state of high alert, prepped to flee or defend ourselves.

Most of us recognize some physical symptoms of stress, but not always as a sign that we've been triggered. It can be even harder when the physiological reactions are less extreme because we often don't associate them with a triggered template—mainly because we might not even remember the template. To help you think about your own reactions, read through this list of other common signs that we've been triggered—our "tells."

Tightness in your chest

Hot feeling in your chest/stomach

Nausea/sick feeling in your stomach area

Change in how fast you're talking

Voice change—higher, lower, or shaky

Forgetting words or details

Sudden headache or pain behind the eyes

Avoiding eye contact

Staring intensely at the other person

Biting your nails

Clenching your hands

Drumming your fingers

Clenching your jaw

Playing with a pen or another small object

Swiveling in your chair

Bouncing your knee

Rocking on your feet

Your body doesn't lie, but your neocortex surely does. It lies through its teeth to try to protect you! Remember, as we said in chapter 2, the good old neocortex is where we form our life strategies, so logically, in wanting to protect us, it can use (deceptive) defense strategies to "help" us feel we are coping. It is the biggest liar on the planet—but well-intentioned.

EXERCISE: KNOW YOUR EARLY WARNING SYSTEM

It can be difficult to recognize our unique early warning system—those physiological responses that are specific to each of us that can tell us we're being triggered. To help, we ask people to do an S+T=B *deep dive* into a recent moment when they know they were triggered because they resorted to a BTL coping strategy.

I.   To do this, first think of a situation. Then, go deep by considering the following questions. What was the situation? See it in your mind. Focus on what was there, the surroundings, the colors, sounds (even smells if you can).

2.   Who was there? What was said? What was the tone?

3.   How did it make you feel?

4.   Consider each part of your body: your hands and feet, your arms and legs, your core, your face and head. What was happening as your emotions began to get stirred up?

5.   Identify how often this happens and in what sorts of situations.

6.   Finally, can you identify how long you've been living like this? Ask yourself, "How old is that reaction?" (This can be a great limbic question to unlock some information for you.)

If you're feeling stuck, keep going deeper into the details of the scene. You may even begin to reexperience the emotions, and if you do, the physiological reactions will crop up. When they do, don't start rationalizing them away. Just feel the sensations and make a note of them. You can practice distancing yourself from the physiology and emotions if they get too strong by saying (out loud if you need to): *Oh, here are these [emotions] and [physical reactions]. They are just happening to me, they are not me. That was then and this is now. I'm going to breathe and turn my mind away from them now.*

The goal with this exercise isn't to relive or work through a triggering situation. It is simply to learn about how your body responds when you are triggered, from the very beginning, so that you're wiser and can see it coming, feel it coming, the next time. Learning to correlate our physiological reactions with templates can help us become more aware so we can better manage our responses.

The power of knowing when we are being triggered is that we then have an opportunity to stop, consider what's happening, and make a better choice about what we'll do next. We can override our helpful neocortex's coping patterns and create new templates for ATL strategies.

2. SBTB: Stop, Breathe, Think, Behave—What to Do in the Moment

The process for *Stop, Breathe, Think, Behave* basically goes like this:

1. **STOP (shift happens!):** literally stop what you're doing. If you are speaking, just finish your sentence and pause. If you're sitting or standing, stop and be aware of yourself, then make a tiny shift in your physical posture (move your weight to the other leg if standing, shift position in the chair). It only needs to be tiny, so others won't even notice, but *you will* because you have given it meaning as part of this process. The outward movement is small, but your thought of what it represents is big!

2. **BREATHE:** take a breath in through your nose and exhale. Some people have asked how they're supposed to do this without stopping the conversation and looking a bit weird to the people around them. It doesn't have to be that dramatic. Simply allow others to talk while you silently breathe

in and out deeply through your nose. Expand your lungs. If you're sitting, sit back, not forward, with a slightly more relaxed or open posture. If you're standing, shift your weight slightly. All of these small intentional changes signal the brain to "calm down." Even taking two deep, intentional breaths makes a significant difference to relax the intensity and reoxygenate your brain, to think more clearly.

3. **THINK:** The Three I's:

   What's my *intention* here?
   What *impression* is that creating to the others?
   What *impact* is my behavior having on others? Be
   intentional (and honest!) about your answers.

Then, consciously focus on a positive word to say to yourself, like "calm," or as we say to ourselves, "Go where love is" (from our One Minute at Base Camp exercise in chapter 1) and feel yourself shift *from* your current physiology *to* a calmer place. Some people even say to themselves, "Go above the line" or focus on specific styles: "Go to compassionate, authentic, relating," and so on. The most effective way is to breathe then shift thinking, breathe then shift thinking.

4. **BEHAVE:** Shift yourself to what ATL thinking or behavior you want to be using right now. Mara uses this process when she feels she's BTL in striving (or when Stephen kindly points it out to her if she's in the thick of it!): she stops, becomes aware of the tension in her body, relaxes her body, breathes, says, "Let go," and shifts up to achieving (as we know, a *much* better way to get something done than striving!).

Simple, right? But sometimes very hard to do in a moment of stress, anger, or anxiety. We've developed a set of tools and tactics to support the SBTB approach.

Stop It—Shift Your Posture, Your Voice, Your Language

Often, when we're being pulled BTL, we may recognize some of our physiological responses, but not all of the ways our body and voice are being affected. Our posture and voice are aspects of our behavior and can send big signals, as we all know—it's not what you say but how you say it. Adjusting our body and voice has an effect on our thinking, too, helping to calm us.

The next time you feel yourself being triggered, try these simple steps to make sure your body and voice are sending ATL signals:

- Relax and drop your shoulders.

- Drop your hands to your side or onto the table or desk, or gently clasp them in front of you (so that you don't point, jab, or make slicing gestures as you speak).

- Relax your face and your throat.

- Smile, feel the muscles change, and also smile with your eyes.

- Consider shifting your voice up or down an octave or up or down in volume, depending on where it tends to go when you are slipping BTL. Some people get loud when they're angry or offended, while others get quiet. Some talk in a lower octave, some higher. Know where you fall and consciously shift away from it.

Breathe It—the Brain Drain

Have there been times when you couldn't focus on what was happening in front of you because you were consumed by the thoughts in your head? Or maybe you're lost for words, or an argument becomes ridiculous just because we can't find the car keys! It can feel like logic and reason have been swept away—and that's *not* so far from the truth!

When we are under stress and being triggered, our brain decides it needs to deal with a threat—real or imagined—constricting the blood vessels in the neocortex, where logic rules and our social behavior is regulated. It redirects blood flow (well, not all of it, otherwise we'd die!) into our limbic system, where our emotions rule, our templates are stored, and the fight-or-flight response begins. Our cognitive processing declines as we shift into "survival mode." Blood flow in the brain shifts to the part that is most responsible for keeping us alive when we're in danger—even if the danger is just emotional danger.

In that moment, when we're under the influence of *brain drain** our insight, our judgment, our objectivity, and our self-awareness are limited. All of these functions occur in the neocortex. Our focus narrows, and we develop a form of tunnel vision (physical and emotional), unable to think about anything much beyond the threat in front of us—the specific thing that is triggering us. If you're in a problem-solving meeting and you suddenly feel threatened by somebody's comment about your idea, your logic and objectivity can go out the window. You can go blank and forget information you know like the back of your hand. Then you start worrying about that, lose your focus, miss what's being said, and worry about that, too! The blood flow goes to the brain's fight or flight center, then your limbic system kicks in, trying to identify past situations when you felt like this. When the limbic system dominates, we have a hard time trading up or overriding old templates (Daniel Goleman originally called it the amygdala hijack). Our cascading responses and physical sensations override everything else. What a mess! And we usually try to deal with it in the worst possible way—by fighting it. Doesn't work, does it?

If our brain does this automatically, how on earth are we supposed to handle it? The answer is ridiculously easy, actually—although hard to remember in the moment. Just stop (because you recognize you've been triggered) and *breathe*. Yes, it's really that simple (and that complex).

When we feel ourselves beginning to obsess over what was just said, or debate the pros and cons of various responses, or any of the signs that we've been triggered, we need to take a couple of deep breaths. It helps reoxygenate the blood and calms the nervous system. Blood vessels expand again as the blood flows back into the logic and clarity center of the brain, and we can think more logically and objectively about the situation and see what's happening for us, physically and emotionally.

When your brain is fully functional and you are aware that you've been triggered, you can think about why you are responding as you are and connect with what's happening in your heart. Then your neocortex can step in and offer effective ATL strategies, instead of the default "let's protect" BTL coping strategies.

Breath is free and it is the best way to access the spirit of love. In fact, "spirit" means breath. When we stop and breathe, not only do we reoxygenate the brain but we shift the spirit in our heart, changing the atmosphere within ourselves and the vibe resonating from us. As Stephen often says, "Every behavior has a spirit on it, every spirit has a behavior on it." With our breath we can transform the energy around us.

Think It—with Your Head and Heart

When he was a young man, Brian, a likable and intelligent entrepreneur, started a business. Unfortunately, his youth and inexperience did not lead to huge success. The business failed when a vendor took advantage of his lack of an attorney when agreeing on the contract. Brian had trusted the vendor because he seemed like a good person, so he hadn't been rigorous in reviewing the contract. That decision almost ruined him financially and emotionally. Even his marriage suffered.

Brian had the resilience to start another business a few years later that proved more successful. By now, he and his wife had two young children and Brian felt the pressure to avoid putting the family in financial danger. His old template from his first business experience kept him from ever considering a partner. He kept his relationships with his vendors transactional and focused on building a self-reliant company. In that role, he could be controlling and striving, believing that "perfect" work from his team and having a hand in every decision would keep his business safe.

When one of his trusted leaders brought a business opportunity to Brian, he felt he should listen. The result was a meeting with a potential partner who could add to their service offering. It really did seem like a wonderful match. The first meeting led to another, and another, and another. Amazingly, despite Brian's feet dragging, there came a day when they had contracts in hand.

Throughout the process, Brian was self-aware enough to recognize why he was resisting. He even said to his wife more than once, "Remember what happened last time." He also knew, though, that this was a great opportunity, he had done rock-solid due diligence, and he genuinely liked their potential partner, Joanna. Joanna knew the industry extremely well, had been in business longer than he had, and was a pleasure to be with. They had had more than one dinner out with their spouses to get to know each other better. Even when having fun, though, Brian would become tense when the topic of the partnership came up.

The final meeting to walk through the contract together arrived, and with each clause Brian's anxiety level was rising. He felt light-headed as they start talking about profit-sharing. His voice became tight, his words clipped. *I can't sign this!* he thought. *What if it fails? What if she gets greedy and tries to screw me over? I can't put my family in danger again!* Then, just as he was about to go ballistic over who got final say on marketing plans, he realized what was happening. He stopped. He breathed. He said to himself, "That was then and this is now. It's different, and it's okay now. Step into courage, man."

An hour later, they were celebrating their new partnership and the great things to come.

What happened between breathing and signing the contract that allowed Brian to take a leap of faith? He answered a basic but essential question: "What is happening in my head and heart?"

Despite his old templates and patterns of behavior, Brian was a different person than he was when he experienced that first business catastrophe. He was now an experienced, knowledgeable businessperson who had done his due diligence. He also was considering partnering with a very different person, somebody who had been in business for years, had a great track record, and had no history of bad business deals that he could find. Brian needed to build faith—in himself through humility and in others through love. But his old attachments were making it difficult.

One way we describe the difference between ATL and BTL behaviors is in terms of the difference between attachment and faith. ATL behaviors are grounded in faith in the product or service, faith in people, faith in the team, faith in supportive family and friends. BTL behaviors are defined by attachments*—to the perfect outcome, to our status, to being right, to the image we project to the world, to being respected, to gaining approval, to maintaining "artificial harmony." We are attached because we believe that those things can prove our worth. When we coach people on how to shift away from one of those coping strategies, we often ask, "What are you attached to in this moment?" For Brian, it was double-edged: protecting himself and his business and proving that he was a savvy businessperson.

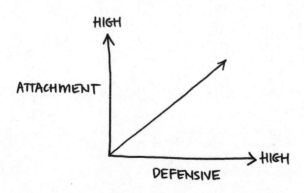

Attachment is insidious because it's easy to justify. We might use positive phrases like "detail oriented," "type A," or "passionate" when we're striving or controlling and are attached to perfect outcomes and the way things are done. We can use phrases like "team player," "polite," and "get along with people" when we're actually attached to how others perceive us, so we dodge risk and are dependent and avoiding.

In the contract meeting, Brian took a first step of faith as he thought through what was happening for him. He thought about what he knew of his potential partner and what he believed her true positive intent was. He then chose to trust Joanna with his "dark secret" of failure. With authenticity, he briefly described what had happened with his first business and how that experience was making it difficult for him to feel comfortable with the new business relationship, causing him to be distrustful and combative. That simple step freed his heart and helped his new partner understand him better. It helped deepen their trust partnership.

Letting go isn't letting go *of* a dream. It's letting go of the *attachment to* the dream. When you let go of attachment, you can rise up and be your best self, and be open to development and ideas. You also open yourself up to an opportunity greater than you expected.

Behave It—Choose a Style and Be That Way

Let's keep this really simple.

Really practical.

Don't overcomplicate it.

Choose an ATL style. Study it, and practice the related how-tos (chapters 7 to 10 are your resource for this).

Think about what you think about—S+T=B.

And behave that way!

Strategic Withdrawal

Sometimes our best option after we've been triggered and think we might end up BTL is to say nothing, to do nothing. In our solve-the-problem, get-it-done-now world, that can also be the most challenging choice. This is even harder if you tend to struggle with behaviors like controlling, striving, approval seeking, or dependent. Strategic withdrawal is not the same thing as avoiding, because we are operating from a place of humility or love rather than a place of fear. We have the wisdom that either we aren't yet in control of our own emotions, or the other person or people are not ready to hear what we might need to share with genuine authenticity and compassion.

If we aren't careful, though, strategic withdrawal might easily send BTL signals, so we need to use the strategy with care. If you're in a group interaction or one-on-one situation, develop go-to language to express your positive intent, such as, "I think this is important and we need to discuss it, but I'm not sure this is the best time. Why don't we take a break and come back to it [tomorrow]?" A phrase like that can help prevent a BTL downward spiral that can be hard to recover from.

Or, as one of Stephen's wonderful PAs used to say, "Time for a walk!"

Of course, if we want to be authentic and achieving, and strengthen our relationship, we have to *actually* come back to it. The danger of strategic withdrawal is that it can turn into avoiding if we let it go and keep letting it go. Be purposeful in making a plan to connect with the person and revisit the conversation at a time when emotions aren't running so high.

3. Plan Your Character—
What to Do in the Future

Tomorrow, next week, or next month, you have something on your calendar that is already creating stress in your life. Maybe it's a

EXERCISE: PLAN YOUR CHARACTER

I.   Identify upcoming trigger situations. Look ahead on your calendar and note any meetings, commitments, events, or interactions that may present triggers for you. Why do you think there is the potential to be triggered?

2.   Identify potential BTL behavior you might encounter from others. What do you think is happening in their hearts and minds that could send them BTL? What is the positive intent within those behaviors?

3.   Identify what templates might be triggered for you. How have those templates caused you to react or behave in the past in similar situations?

4.   Trade up! How would you *like* to respond or behave instead to be more aligned with your personal values and to be more effective?

5.   Swap your templates: What more effective template—one formed from *positive* experiences and the gold at the core of your heart—could you commit to use in the situation?

meeting with your team and your boss about a project that isn't going according to plan. Or maybe it's a difficult performance meeting with a team member, or houseguests who have overstayed their welcome. Whatever it is, what you know is that something about the situation will trigger you, and your brain will pull up one of those old templates that it has been pulling up in similar situations for years. And things will invariably not end well!

We spend time thinking about the *content* of our future interactions. We prepare agendas and slides and reports for meetings. We spend time perfecting the vacation stories we'll tell at the next social event. Yet, we often don't think to plan our *character*—our heart attitudes, thinking, and behavior—even when we know we may be triggered. It's a powerful way to help ourselves stay ATL and feel good about our interactions, especially the tough ones, and it requires just five simple steps. Taking just a few minutes to run through this exercise can help you defuse below-the-line situations before they even begin.

The X Factor—Shifting Your Heart Attitude

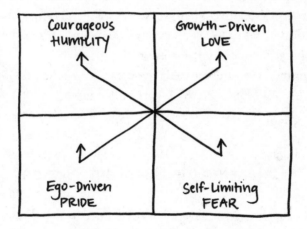

Brian, and every other person we've written about in this book, shifted their behavior by shifting their heart attitude.

Growth-driven love overcomes ego-driven pride. Courageous humility overcomes self-limiting fear. We call this the *X factor*: we often see people who score high in growth-driven love score lower in ego-driven pride, and people who score high in courageous humility score lower in self-limiting fear.

As you know by now, the need to prove ourselves (through pride-driven behavior) can make us obsessed with self-promoting. If we focus on the growth-driven love in our hearts while achieving results, we become more engaged with genuinely serving, respecting, and honoring others, overcoming that self-obsession. The need to protect ourselves (through fear-driven behavior) can make us obsessed with self-protecting. (The paradox is that when we spend our corporate life self-protecting to retain our job, when there's a restructure we may be first to lose our job!) If we grow and focus on the courageous humility in our hearts, we become accepting of ourselves for the wonderful forces of good we are, even with our opportunities for growth. Acceptance overcomes the obsession with approval.

So what does this mean in the moment, when we're thinking about what's happening for us? The simplest first step is to focus on growth-driven love when we know we're operating out of ego-driven pride, and to focus on courageous humility when we know we're operating out of self-limiting fear.

It's an increase to decrease, as you can see in the previous illustration. As ATL pulls up, BTL gets pulled down.

Balancing the Quadrant "Scales"

In a set of scales, if one side is heavily loaded, it tips down—it's the same with the Heartstyles quadrants. One of the biggest challenges we see when leaders learn Heartstyles is they can think it's all about *overly nice* "tree hugging," and they forget about achieving results. The scales can become uneven, out of balance, and thus ineffective. If there are *high* growing others scores and *low* personal growth scores, the scales will tip *down* and the behavior *will look and feel like* self-protecting (*too* passive, *too* nice), and results will suffer. Likewise, if there are *high* personal growth scores and *low* growing others scores, the scales will tip *down* and the behavior *will look and feel like* self-promoting (too aggressive, too me instead of we), and culture will suffer. True north is a *balance* of *both* the above the line quadrants on the scales.

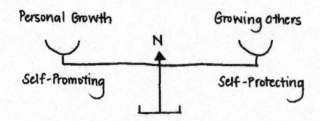

The Behavior Debrief: Use What You've Learned to Do Better Next Time

When was the last time you can remember being pulled BTL? Afterward, what did you do?

What we often do is obsess about details of the situation that had little to do with the outcome—often minor behaviors of other people or facets of the situation beyond our control—and then justify our behavior based on this "introspection." We can spend a lot of time thinking and talking about why things happen without getting to any actual truths—we just get stuck going around in circles. A bit like a hamster in a wheel, running in place and not moving forward.

That is the basis of our Behavior Debrief, which is a quick analysis of S+T=B, working backward to collect your facts and opinions, and then a future-focused conclusion. We recommend writing your debrief notes down—in whatever format works for you—rather than just thinking through the questions, so that you can refer back the next time you need to plan your character for a similar situation. We also recommend doing this once a week for a while as you start working to grow away from a certain BTL style or strengthen certain ATL styles.

The process of change takes time and a willingness to keep going

EXERCISE: BEHAVIOR DEBRIEF

S: What was the situation? What about the situation was similar to others in which I have behaved the same way?

B: What was my behavior? Was the outcome effective or ineffective?

T: What was happening for me? What triggered me? What template was in play? Was I attached to an outcome or aspect of the situation? Where was my heart?

Conclusion: What is *one thing* I can do differently next time? How can I implement that—what's my *first step*?

through the awkward early stages. Many of us believe a myth that has lingered since the 1960s: it takes twenty-one days to change a habit or to develop a new one. Actually, a study published in the *European Journal of Social Psychology* found that on average it takes sixty-six days for a habit to become ingrained, for new neural pathways to stick and override old ones. The more effort the new habit required, the longer it took to establish. A regular exercise routine might take longer to establish than daily journaling (although for some of us, the exercise would be easier than the journaling!). Our brains are hardwired to take shortcuts and to do what comes naturally to us, fighting against the effort to change. Time and consistency are key.

As you begin to use the strategies we've shared for shifting your heart, thinking, and behavior ATL, develop compassion for yourself. Life is complex, and we're constantly being challenged by new responsibilities, new relationships, and new struggles that can pull us south BTL. Focus on tackling that thing that has you most confused or distressed right now, and keep moving forward. In the coming chapters, we'll share more helpful strategies that you can try, depending on what has you feeling stuck right now. Next month or next year, when you're tackling something new, come back to these chapters to refresh your approach. Remember the approaches work for all struggles, so you just apply the approach to the newest issue!

And in the meantime, trust the process (TTP!) and have patience. Life is full of opportunities to grow. Those opportunities create a wonderful journey of becoming who we know we can be.

To equip you with more know-how to live and lead ATL, the ATL styles are summarized by: Know Who I Am (Authentic and Transforming), Know Where I'm Going (Reliable and Achieving), Connecting with Others (Relating and Encouraging), Growing with Others (Developing and Compassionate)—our next four chapters.

CHAPTER 7

Know Who I Am—
Authentic and Transforming

Be yourself. Everyone else is taken.

—attributed to Oscar Wilde (and a few other people!)

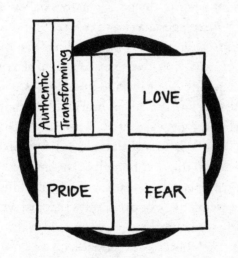

Deep within us there is a cry to be our true self. To not be caught in the *fortress of fear* or the *prison of pride,* but to be *free to be me.* If we're in an environment, culture, or even a relationship that doesn't allow us to be authentic, we can take the opportunity to be bold in changing the situation as much as we can influence it. Maybe you know somebody who is secure in themselves, who really knows who they are and operates with courage and integrity, based on what they know they value. They are honest with themselves and with others, and that vulnerability usually paves the way for others to be real, too. They seek out change and growth and aren't defeated by honest feedback—in fact, they go looking for it. This is what life looks like when you are strong in authentic and transforming styles.

The people we respect because they seem to stay cool, calm, in command of themselves, and aligned with their values when surrounded by the common chaotic moments of life are simply operating out of the core of their heart and the best aspects of their character.

Henry was a product development manager, leading the development of a groundbreaking new product in the engineering industry. As Henry's team was overwhelmed by the demands of managing this big venture along with many smaller ones, the CEO of the company hired an executive to oversee all projects. The new leader looked great on paper, spoke the lingo, and seemed like a good fit. Unfortunately, that was not how it turned out.

Month after month went by, and every milestone for Henry's project was missed. The CEO kept hearing, "It's those guys in development" from the new executive. The truth? The man was in over his head, wasn't as knowledgeable as he reported himself to be, and so felt he had to prove himself as the smartest person in the room. He was competitive, sarcastic, and controlling, demanding that the team develop the product in ways that just weren't going to work. When Henry and his colleagues would make other suggestions based on their collective experience, he would override them.

So what did Henry do? Nothing. Henry was entrenched in fear, so afraid of rocking the boat that he deferred to somebody with less expertise—even though he knew his knowledge was correct. To feel safe, Henry depended on his boss to tell him what to do, and so he kept quiet.

You might guess what Henry's Indicator showed when he completed it: high in approval-seeking, dependent, and avoiding behaviors. Faced with the truth of what was driving his behavior—

fear—Henry had a breakthrough. In a session with us, he explained that he was the same way at home with family and friends. He kept smoothing over challenges in his marriage and strife with family members rather than dealing with them directly. Anxiety, depression, and insomnia were defining his life—they felt normal to him. Yet that was not the way he wanted to live. "I'm creating a lifestyle of lies," he suddenly announced in one conversation. (We still use that phrase today!) "I'm actually lying to myself. I'm lying to others. I don't want to be like this anymore."

It was a tough day for Henry when he finally walked into his boss's office to have some care-frontation. Unfortunately, it did not go well. Nor did Henry's next two attempts to discuss the situation. His boss was consistently unwilling to listen to Henry's concerns and appeared aloof and distracted—even answering text messages while Henry was speaking to him. Henry left the third meeting feeling saddened and frustrated—and knowing he had to do something more about the situation, because it was his responsibility. He realized that if he didn't, he wouldn't be true to himself, his team, the project, or the company's investment.

Two days of mulling led to him stepping into the courage to be authentic, knowing he had to be the one to change the situation, and he organized an appointment with the CEO. The next day Henry, a little apprehensive, walked into the CEO's office and said, "Can I be authentic with you? I need to explain what's really going on"—a tough moment, but it might have been one of the most important days of his life. It was the day he began to trade up, away from fear and into authentic and transforming. It wasn't an easy journey, but today Henry is one of the most straightforward and honest people you'll meet—with himself and others. He's a man of integrity in difficult moments. He's always looking for opportunities to develop, to learn more about himself. He's also less anxious and more confident as a person, a partner, a parent, a leader. A man who used to live in fear now helps others find the courage to be their true selves. That's *real* truth, not lies.

Recap: Authentic and Transforming

If you scan the Heartstyles model from left to right, authentic and transforming are the first two ATL styles you see. Why? *Together they catapult our self-awareness.* **That's why these two styles are next to each other on the Indicator**—they're the Heart in the Heart + Smart of the humility personal growth quadrant. Reliable and achieving are the Smart part (next chapter). When we are truly authentic (real and transparent) and transforming (continually learning and developing), we're honest not just with the world outside of us but also with ourselves. By overcoming fear in particular, we develop the *courage to acknowledge our opportunities for growth without it affecting our sense of who we are.*

Operating in authentic and transforming styles means being comfortable openly discussing what we discover about our strengths and weaknesses, receiving feedback, and delighting in finding ways to improve. The Indicator includes questions about the quality of participants' lives, and we see that transforming behaviors are often strong in people who report a high level of focus on personal development, but also high levels of happiness, health, and personal and professional effectiveness. Authentic people are often fearlessly open-minded and trustworthy. "Integrity" is one of the first words you would use to describe them. Their authenticity creates a safe place for others to be vulnerable and open, too, and that creates better relationships.

It's not surprising that in all of the conversations we've had with CEOs and other executives about the top attributes they want in their new hires, authentic and transforming are at the top. Even a *Forbes* article that captured the thoughts of CEOs from some of the best companies to work for reflected that thinking. Phrases like "growth mind-set," "curiosity," "humility," "lives values," and "sees failures as learning opportunities" came up again and again. Developing this thinking and these heart attitudes allows us to break the hold of fear in our lives, to say enough is enough! True character is all about staying ATL in the face of criticism, injustice, disappointment, and betrayal.

All of us, though, can amplify our authentic and transforming behaviors through the six tips we outline here.

1. Turn a Breakdown into a Breakthrough

Identifying the specific kinds of BTL behaviors that we incline toward is the first step in trading up to authentic and transforming behaviors.

In most social situations, you would barely notice Elena. What struck us, the first few times we met her, was how quickly she removed herself, escaping to the kitchen or the bathroom. She rarely looked people in the eye. She was difficult to get to know.

On a retreat for a group of leaders and their spouses, she sat quietly through day one, but on the second day, one activity in particular, used to tap into the limbic system and heart, drew her out. As she explored a challenging situation in her life, and what was driving her behavior and that of others, she began to lose her composure. Her fear began to bubble up and, unable to go forward or ignore what she was uncovering, she stopped and did the one thing that was hardest for her. She drew attention to herself. Overcoming her evident fear of what other people thought about her, Elena stood up, said, "I can't do this," burst into tears, and hurried out of the meeting room. That afternoon and the next day, she didn't return, staying locked in her room.

Elena's fear had her trapped in a mind-set, heart attitude, and body that were not the truth of who she was. She was obese and facing some critical health issues; her body issues both stemmed from and fed her feelings of low self-worth and fear of rejection and criticism. She was consumed by her fear of people's judgment, by her worry over what people thought of her, because she never felt good enough, pretty enough, or smart enough. The grip was so strong that she was even triggered by weight control programs or professionals, assuming coaches and participants were judging her for her "failures"—and that kept her from seeking help.

But sometimes our break*down* can be our break*through*. In that moment in that room, facing a breakthrough to what was happening for her on a deeper level, she had no choice but to be authentic. It may not have been highly productive, but it was honest and it was stronger than the fear of what other people thought. The day she returned to our workshop, we noted there was something different about her— her courage was evident as she set out to explain what had happened for her. In a private conversation, we asked her, "What if your desire to improve yourself and your courage to make the changes you've wanted to make for so many years are stronger than the fear?"

Elena had proven to herself that she had the courage to break out of the fortress of fear—what she thought people would think of her. Her first step allowed her to take the next one. She spent the day asking, *What's the gold within my heart? How am I authentic and courageous? How am I loving? How am I loved?*

Over the coming months and years, Elena worked to trade up to authentic and transforming, away from avoiding, dependent, and approval seeking. Patiently and persistently she worked on identifying the voids and wounds of her heart and healing them. She said to us at one point, "I've got a woman trapped inside of me, and that woman has a voice. I've been so afraid all my life to share it. But I'm going to find her."

Elena joined a program to address her chief challenge—using food as a form of comfort to overcome her sense of failure. She got a coach. She became an engaged part of a community of people, encouraged by them, encouraging them, and being vulnerable with them. She began to come out of her shell and share with the world the beautiful woman that she was discovering within, without fear of judgment. She came to enjoy food again, because it wasn't so laden with judgment and fear of failure. Day by day, month by month, she got healthier. Over the next two years, Elena lost one hundred pounds. Astoundingly, the person who could barely handle social contact became an authentic and compelling speaker at women's events, standing on stage in front of hundreds of people. At the very first event, we were privileged to see her receive a roaring standing ovation!

Elena's lifelong issues with self-worth haven't completely disappeared, but today she can talk about them with self-compassion, and that has kept her heart and her thinking ATL. Recently, she posted an old photo of herself with the words "Struggling today with self-image. Posting this to remind myself how far I've come and that I am strong."

Each one of us is just as strong as Elena, just as capable of developing our character. The world teaches us to mask vulnerabilities. It might take a tough moment of admitting to ourselves that thing we've been afraid to face, of removing one of the masks we wear to reveal the true person behind it. As we wrote in chapter 1, fear leads to pride leads to denial. Those breakthrough moments of self-aware honesty free us from the denial that keeps us trapped BTL. Like Elena, each of us can come through a breakthrough moment, no matter how painful, and apply ourselves with patience and persistence to our transformation. Like Elena, we can gather around us the kind of people who will support our journey. And like Elena, we may well find that the journey leads us to so many kinds of transformation: personally, professionally, physically, emotionally, and spiritually.

The most courageous, humble thing you can do in your life is to be honest with yourself, and then work to transform into the person you know you want to and can be. Not many people go on that journey. The battle is so worth the security of knowing who you are and accepting that whole, wonderful person who is an AND.

2. Trade Up from Insecurity

In our research, statistically the strongest inverse relationship with authentic and transforming behaviors is with the fear-based behaviors of easily offended and avoiding. As you've learned, the more insecure and fearful we are—feeling not good enough, perceiving correction or constructive feedback as rejection, avoiding our development opportunities—the more we will resort to BTL coping strategies. When we can accept who we are and have confidence in that person, rather than expending our mental focus and emotional energy on trying to self-protect or self-promote to feel okay, it frees us from the anxiety that comes with fear of judgment or craving approval to build our sense of worth.

Growing in authentic and transforming behaviors also helps counter sarcastic and competitive tendencies. When we are sarcastic, we can convince ourselves that we are being authentic, but with humor. In fact, we're often avoiding saying what we truly think or feel—it is an aggressive form of avoidance rather than honest authenticity. That's a tough thing to hear if sarcasm is your go-to style. Sometimes we refer to sarcastic as "starcastic" because it can be about "look at me"—another form of competitive and approval seeking. Here's your moment to really look at yourself: Is your sarcasm about being authentic or is it, really, about showing how clever you are? Ouch! Remember *we are all* an AND!

With competitive, we can convince ourselves that we're trying to improve or grow, but really our focus is on *being better than others* rather than *bettering ourselves*. We focus on superficial or outward forms of "improvement," such as trophies and promotions and money, rather than personal growth, especially character growth, that can bring us greater contentment, confidence, and a sense of worth. It's about *showing* others and ourselves how adept we are, through tangible external measures. The extreme version of this is when people cheat to win—for example, in professional sports. Mountaineers are not exempt, we're sorry to say: climbers have reported thefts of oxygen bottles from the high camps on Mount Everest, the ultimate act of rivalry.

3. Find Out Where You Are Strong

Following the bread-crumb path of our unique thinking and heart attitudes is how we get to breakthrough moments that can kick off our journey of knowing who we are, or change the shape of our path. Take a few minutes to consider the following five questions. Use them to guide you in examining where and why you are authentic and transforming, and what may be keeping you from those behaviors in other situations. S+T=B again!

Look in the mirror:

- When have you found courage, humility, and "gold" in your character that enabled you to be authentic with yourself and others? What is a common thread in those situations?

- Where in your life have you been transforming? What drove you to pursue that personal growth?

- In what situations would you like to bring more of your true self, with honesty and integrity, to a situation, a relationship, or an aspect of your life?

- What is a personal transformation or growth effort you've been struggling with? What aspect of it seems most challenging?

- Reviewing the work you've already done to examine your triggers, templates, truths, and voids, wounds, and inner vows, what specifically do you think is blocking you from being more authentic and transforming in certain situations?

4. Anchor Yourself to Your Values

Jeff sat in the back corner of the conference room at a leadership team off-site. He was leaning away from us, his arms crossed, a slight scowl on his face, watching as we introduced ourselves and the program we were about to embark on. When it came time for introductions, he didn't hold back: "I'm sorry, but I have no idea why we're here. Our business is going downhill, we've each got a hundred e-mails in our inbox right now, and we're spending time and money on *culture*? On *love* and *humility*? You've got to be kidding!" He was there under sufferance, and as the CEO, was wondering how he had agreed to this "waste of time and resources."

The culture of fear and pride at Jeff's company was bolstered by his own sarcastic and controlling behaviors. He was smart and very good at what he did, but he had a habit of shredding people and ideas with cutting wit. People were afraid of him. And his passion for solving the issues of the business was a good mask for his inability to address his own.

The turning point was seeing his Indicator Benchmark versus his Self and Others results, plus the aggregated results of his team. As we discussed the effects of sarcasm and the value of authenticity, we used the word "integrity"—speaking our truth, from the heart, with integrity is the essence of authentic behaviors. That word resonated with Jeff; he believed in the importance of integrity; he had claimed it as a personal value he wanted to uphold in his life. When he realized how far his sarcastic behavior was from true integrity, it was jarring. He was able to see how his behavior with his colleagues stemmed from fear and pride, how it was generating fear and pride in response, and how that was impacting the company culture—straight from the top!

Jeff's personal transformation took time. To his credit, he took courage with both hands and ran toward his own development. Jeff asked his team and colleagues to call him out when he was being sarcastic or using competitive "us and them" language. He had other people run meetings so that he wouldn't dominate. Mostly he worked on making people feel safe around him—and because of it, he became a more effective leader and coach. Because Jeff took the first step, others could take the second. Over time, people's trust in him grew. And over the next months and years, he became one of the most authentic and transforming leaders we've worked with—personally, with his team, and with his company.

When we define ourselves by how we live our values, our self-worth becomes independent from our behaviors and outcomes, and that allows us to be more authentic because we are less focused on protecting ourselves from criticism or proving ourselves. When Jeff began to shift his heart attitude and became less attached to being right and being seen as "smart," his authentic self—his high values, belief in fairness, and commitment to helping others achieve their best— began to shine through. His joy for his work expanded and his relationships with colleagues and team members became more fulfilling.

So how can we all focus our attention on living our values? Well, not to sound sarcastic, but it helps to know what values we actually want to live by. Jeff did, and that self-knowledge was the source of a breakthrough moment, a higher level of self-awareness. There are a number of good online tools that will help you sift through many kinds of values and name the ones that are most meaningful for you. Once we know our values, we need a way to keep them top of mind in daily life.

Values Discovery Process

An easy-to-remember tip we often recommend to people who struggle to keep their values top of mind is to build an acronym. Begin with a word that encapsulates who you are or your highest value. You might even use one of your names. Now, turn that word into an acronym, using words or phrases that capture your values. For instance, Stephen's is based on our surname:

K: kindness

L: love

E: economy of enough

M: ministry in service of others

I: integrity

C: courageous in character

H: humble heart

When you feel yourself being triggered in a way that might lead you away from authenticity, say your word to yourself and pick the letter and value you think you need to bring to this situation to stay ATL.

Plan to Live Your Values

Our values help define who we are, so understanding them with clarity helps us know ourselves better. Just like planning your character (described in the last chapter), or as a part of that same process, we can plan ways to live our values and beliefs that help us stay authentic in tough situations or help us grow our authenticity. You can choose one behavior related to each of your values that improves how you live it, strengthening your daily, practical reliance on that value. You might also examine which values you struggle to live, or which values tend to fall by the wayside when you're triggered in specific situations. Based on that assessment, you can choose one or two behaviors to rely on in those moments that will help you behave with integrity more consistently.

One of our clients was a hard-driven director of operations, a fairly ATL leader, and becoming more authentic every day. His real challenge was at home. Jim shared with us that he wasn't the father or husband he wanted to be or knew he could be. His evenings were consumed with work. Jim's own striving and controlling kept him preoccupied with e-mail and text messages from the moment he walked in his front door each evening. As soon as he would answer a call, he noticed, his kids would start acting up, working hard to get his attention. He knew the mobile device was ruining his evenings with his family and creating tension, so we helped him develop a plan to align his behavior with how much he valued family in his life.

Each night, as Jim approached his house, he would pull over to the side of the road and address any truly urgent messages. He would then switch his phone to airplane mode so that he wouldn't be tempted by buzzing and pinging. For the first hour at home, he would focus entirely on his family. After giving his family his full attention, if he needed to communicate via his device, he would remove himself from the family scene, letting his family know what he was doing, and go to another room and deal with it. This allowed Jim to stay fully present and focused when he was with his family. If he knew that he was expecting something urgent that evening, he would frontload it with his family: "Dad has a call coming tonight that I will have to take."

The results at home were powerfully positive, but Jim also discovered that setting boundaries with his team in the evenings, letting them know that he would only respond to the most urgent issues, empowered them to make decisions and created better work/life balance all round. Stress levels fell and autonomy and satisfaction rose.

5. Truth Teller—Be One, Find One

Not long ago, we went on a trip with two very good friends, a couple. As part of the trip, we were planning a climb up a rock face in the Italian Alps. Tom had been excited about the prospect, looking forward to the challenge it would be for him. In fact, he instigated the idea.

After we chose the spot—a climb Stephen had done before—and talked about it over dinner, Tom started to do his research. He found photos and videos online and accounts from others.

The day before the climb, he told Stephen, "I'm sorry, I just can't do it. When I think about it, I feel sick. You know I'm afraid of heights, and I have been working on challenging myself, but I think this climb is just too much."

This might seem like letting fear limit one's life. In fact, it was the *opposite* for Tom. Years ago his go-to BTL strategies were competitive and approval seeking. When we are entrenched in approval-seeking behavior, being authentic is incredibly difficult. Add ego-driven competitive behavior, in which comparison and being better than others matters more than our own growth, and Tom's true self—his values, his deep wisdom, the gold within his heart—had little opportunity to see the light of day. Those behaviors are the reason that Tom, for the first ten years of our friendship, would not venture into outdoor activities with us. He thought he couldn't compete, so why risk (he thought) humiliation?

If those behaviors still had Tom in their grip, how would he have handled the situation? He might have waited for the day of the climb and pretended to be sick. He might have attempted the climb and been forced back down, resulting in a feeling of humiliation that might cause him to withdraw from our friendship. What also made the situation a growth opportunity was that his wife, Liz, still wanted to do the climb. The idea that she could handle a climb that he couldn't might have ruined the entire trip—*if* he hadn't been working on strengthening his authentic and transforming thinking and behaviors for some time.

With a strong sense of self and personal acceptance of his capabilities, and a desire for everybody to have a wonderful trip, Tom was able to truthfully share his fears, as well as his desire to push himself in some way, even if it wasn't on the rock face. We decided that Liz and Mara would still do the climb while he and Stephen trekked up the (very steep!) backside of the mountain. We would meet at the top. Tom may have doubted that decision or felt a strong tug BTL when he saw Carlo, the gregarious Italian mountain guide who would be spending the day attached to his wife! Instead, Tom was genuinely encouraging and happy for Liz to tackle a new challenge. Even after the fourth picture the girls and Carlo took together—before we even left the parking lot! The celebration we had at the summit was a true celebration for everybody—as it turned out, the trek was just as hard physically as the rock face!—and we all helped each other on the long trek back down. Now we have a great story to share about our travels together, rather than an experience that everybody avoids talking about, thanks to Tom's authenticity.

Being a truth teller can be challenging and takes awareness and courage. As with everything, we need to pay attention in situations when we may be triggered to hide our true feelings, to avoid sharing an opinion, or to mask our fear of failure by denying ourselves an opportunity for growth. That's why we need others to support us along the way!

Ask Someone to Be Your Truth Teller

When we coach someone, we like to observe them in their "natural habitat." Andrew had some habits that made him come across as overly intense, even aggressive. These communication habits were the lingering effects of Andrew's inner drive to achieve results, but they came across as BTL. In meetings when he was passionate about an innovation or worried about company revenue, he would lean forward in his chair, his expression would become a little severe, he might use words like "must" and "absolutely," and he would make harsh chopping and pointing gestures with his hands. Even when he felt he was operating ATL, these behaviors sometimes made others perceive him as BTL. The challenge for Andrew was that these were habitual behaviors that were very often unconscious. He needed a truth teller.

We suggested he ask his assistant at the time to signal him, subtly, in meetings when these behaviors would start to appear: she would catch his eye and touch her earring. Nobody ever caught on, but Andrew knew to relax his posture and expression, to put his hands on the table, lean back in his chair, and to adjust his voice tone. Slowly, those became his new habits—and people's responses to his ideas and feedback shifted accordingly. Without a truth teller helping him in "live" moments with immediate feedback, that transformation would have been more difficult and taken longer.

In Jeff's story in the previous section on values, you may have noticed a critical sentence in how he became the leader he is today: *Jeff asked his senior colleagues to call him out when he was being sarcastic or using competitive "us and them" language.* Self-awareness cannot be a purely internal effort. The brain is too good at pushing "my truth" above "the truth." Otherwise, sportspeople wouldn't need a coach! We need outside feedback to understand when we are at our best and when we aren't. Only with a broader perception of our behavior can we be effectively transforming.

Try to find a truth teller and give them permission to be honest with an objective opinion, and not just agree with your perspective or avoid harder truths. You want someone who is authentic and compassionate—who can share their thoughts in a way that aligns with what you need in the moment. Of course, that advice goes both ways. If you want the help of an authentic truth teller, it requires you being prepared to be a good "truth listener." Your heart attitude and thinking need to be in courageous humility. If you struggle to handle feedback well, if your coping strategies are dominated by competitive or easily offended behaviors, you may need to do some work to prepare yourself for what your truth teller will share. Remember the SARAH process for managing your inner and outer responses to constructive feedback (it's in your Personal Development Guide).

A last bit of advice: you may find another truth teller in the least obvious place. A good example is those high-level business leaders who look outside the C-suite, spending time and energy seeking the opinions and perspectives of people on the front lines of their company—people who serve customers, make or design products, deal with client challenges, and more. Warren Buffett once tweeted the attributes of the smartest people he knows (none had anything to do with traditional notions of intelligence, by the way). His last point was, "Seek to understand every perspective on a topic." That's tough to do unless you ask people outside your immediate circle. It's true for leaders and organizations, but it's also true for each of us in daily life.

How to Get Good Feedback from Your Truth Teller

Here's a simple approach to gathering helpful feedback. First, ask your truth teller to observe you in specific situations or with a focus on specific behaviors. Then, ask these four questions:

1. What did I do well—praises?

2. What could I have done better—redirects?

3. Did I miss any opportunities to practice an ATL behavior?

4. What's one suggestion you have to help me improve?

6. Being Authentic—In-the-Moment Tools

A friend of ours is a captain with a very reputable airline; he's one of their best, flying the A380 for international flights out of Sydney. On one of those flights across the Pacific, he and his copilot were alerted to a minor technical problem by the infamous flashing red light. They reviewed procedure and conferred with air traffic control and finally made a tough decision. They had more than eight hours ahead of them but less than four hours behind, so for safety they turned the plane around.

Imagine how you might feel if you were a passenger on that flight. First, frightened. A plane turning around four hours into a transpacific flight is no small thing. Second, for some, frustrated, annoyed, even angry. After announcing the decision over the intercom and explaining the reasoning, in his calm voice, the captain began hearing from the cabin crew that people were frustrated, some aggressively so. Leaving the plane in the hands of the experienced copilot for a bit, he left the cockpit and began walking the aisles of the double-decker aircraft. He answered questions as transparently as he could, he patiently explained the decision again and again, and with calm command, he asked passengers not to take their frustrations out on the members of the cabin crew, as the decision had nothing to do with them and their goal was to make the rest of the trip as comfortable as possible. Being this authentic—stepping out of the figurative cockpit, to come out from behind the masks we wear—takes practice and courage.

"Going Public"—It's Not as Scary as You Think

A recurring theme in behavior: people who want to transform don't try to hide their need or desire for growth. Admitting a development need is never as traumatic as you think it will be, especially because we can almost guarantee that the people who know you best already see it. We even talk about "airing our dirty laundry" when speaking about our development needs! Why do we call it "dirty"? When you're intentional about being open and vulnerable with those close to you and giving them permission to coach you, it increases your commitment to cleaning out BTL habits and creating better ones.

Inspired by Brené Brown's exceptional work on vulnerability, a team of psychologists decided to examine an intriguing mismatch, which Brown described in *Daring Greatly*: "We love seeing raw truth and openness in other people, but we are afraid to let them see it in us." If we all want honesty, why do we fear being judged for it? The researchers asked hundreds of participants to respond to scenarios in which somebody else was practicing intentional vulnerability and imagine scenarios in which they did the same themselves. Overwhelmingly, the participants perceived their own vulnerability less positively than the vulnerability of others. The researchers call this the "beautiful mess effect" and concluded, "Even when examples of showing vulnerability might sometimes feel more like weakness from the inside, our findings indicate, that, to others, these acts might look more like courage from the outside." And when others see that courage, they trust us more, are more likely to forgive us, and feel safer with us.

Going public doesn't have to be an announcement from the stage to an auditorium full of strangers. It can be as simple as sharing a growth opportunity with your family, your close friends, your team, or close colleagues. Consider what you want to share, check that your thinking and heart attitude is firmly settled in humility-driven authentic and transforming, and then be honest and transparent. The rewards may surprise you.

Identify Your Fear or Your Attachment

When you find yourself holding back or hiding behind a mask or an untruth, ask yourself these three questions:

1. What am I afraid might happen if I were honest, vulnerable, or transparent right now? Or, what am I trying to achieve by *not* being authentic (what am I attached to)?

2. How will I feel if I don't share my true opinions, feelings, or ideas?

3. If I weren't afraid or uncomfortable, what would I most like to say right now?

Practice Language that Encourages Authenticity

Give yourself the gift of some go-to language you can rely on to help shape your authentic responses. You'll feel more confident and less awkward if you have some phrases to use when you're feeling tugged BTL. While there are no hard and fast rules, and we're not suggesting that you become an "authenticity robot," we do want to share some phrases we often recommend as starters.

- I hadn't thought about it that way.

- That's a helpful perspective.

- I wasn't aware you saw it that way (or felt that way).

- I appreciate your candor.

- I think that idea has value.

You can follow any of these with a transition into what you honestly want to share, with compassion, with variations on the following phrases.

- I see it differently . . .

- Can I be authentic with you . . .

- My perspective is different . . .

- I'd like to share how I feel about the situation . . .

- My idea/thinking follows a different approach . . .

You can set boundaries in conversations, to keep differing opinions from triggering a series of BTL responses and counter-responses, by simply acknowledging the difference between you and that it doesn't reflect poorly on either of you.

- It's okay if we don't agree.

- We may see things differently, and that's okay.

- Our opinions are equally valuable.

If you're in a conversation that is heating up, remember the most important strategy we shared in the previous chapter: *Stop. Breathe. Think. Behave.* Focus on the idea that differing feelings, opinions, or perspectives are not a form of judgment; they are simply different. Let go of the need to be "right" while remaining authentic.

Open yourself to other people's perspectives and what may be happening for them in the moment. You may be surprised by what you discover.

How do you transform your heart, your thinking, your behavior, and thus your life? By seeing the world as full of opportunities to be your true, authentic self and to continue growing into your best self. Each of the strategies we've shared has the potential to help you shift your thinking and your heart attitude, but to do so, you have to go for it! On the surface, many of them may seem easy, and they will be if you take a surface approach. But if you use them with the intention of actually living more consistently ATL, strong in authenticity and dedicated to transformation, there will be some hard work required—but the benefits will be life-transforming. You will discover the power of being your whole authentic self, and then stretching and strengthening that person to be everything you can be. That amazing person then has the power to identify a direction and a purpose, and go out and change the world.

CHAPTER 8

Know Where I'm Going— Reliable and Achieving

Without continual growth and progress, such words as improvement, achievement, and success have no meaning.

—*Benjamin Franklin*

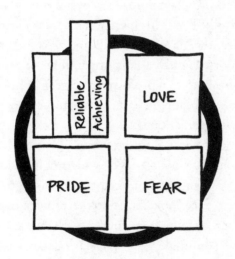

I t is through reliable and achieving behaviors that we create purposeful direction in our lives. This is the Smart of Heart + Smart— the task-focused styles of the personal growth quadrant. Both reliable and achieving work together to set and maintain direction in life—it is purpose-driven, knowing where we're going. It's not a static thing, a single point on a map of the future that we spend our time and energy working toward. It changes, just as life changes.

People who are high achievers often have made the effort to define an inner purpose that guides them during those times of change so that they can course-correct new directions. A connection with a purpose bigger than themselves can also steady them during trials and tribulations. It helps them stay tapped into their ability to make a difference in the lives of others.

Alex had come to Australia with his family to escape the violence in his native country, wanting his children to experience a safer life. In his home country, he had been a senior leader with a property management firm. We didn't know this about him when we first stopped to introduce ourselves, because he spoke little English and he was busy cleaning the common areas of the apartment complex we had recently moved into.

What we noticed right away about Alex was his positivity, his friendliness, and how he approached every task with enthusiasm and care. It was like it was his own home he was cleaning, whether he was polishing the glass in the building elevators, raking leaves, or mopping floors. He had a purpose guiding him steadily on: to work his way back up to a professional career in his new country, providing a good life for his family, whatever it might take. His determination was remarkable.

Despite working three cleaning jobs and having little free time, Alex went to night school to develop his English skills and made quick progress. When he mentioned that he was good with numbers, we suggested he look into bookkeeping or accounting. He quickly found an accountancy course, gathered recommendation letters, and was admitted. We might have been encouraging him, but Alex was inspiring us!

In record time, and while continuing to work long hours, he finished the course. Time and again, he used his goal to inspire himself when he was exhausted.

One year after we met him, Alex got a job in the accounting department of the same property management company that had been employing him as a cleaner. Within two years, he was their senior accountant. Purpose-driven!

Recap: Reliable and Achieving

Stories like Alex's are inspiring. Do you have an equally powerful story of working toward a goal with diligence and excellence until you achieved it? When we know where we are going—when we have a clear sense of purpose and what it will take to move steadily and consistently toward that purpose—we find a special kind of confidence. **That's why reliable and achieving are next to each other on the Indicator.** It's about driving results and getting things done—*healthy* competition. It's also about the bigger picture and our sense of direction in life.

From reading your own Indicator, you'll know that achieving behavior focuses on the vision, purpose, and strategy to get things done with excellence, not perfection. It's about both our direction in life and simply being proactive and getting things done. Based on Indicator results from more than 100,000 people, achieving is a behavior people frequently aspire to, yet one they score themselves fairly low in. It is also the behavior *most correlated to personal effectiveness*—the ability to achieve goals small and large.

Reliable behavior is about honoring others by being dependable and keeping commitments and promises. When we're reliable, we *consistently* meet our commitments, and we do so because we understand the value of discipline and persistence. That discipline guides us to say no when it's necessary, to make effective use of our time, to honor other people's time, and to set and hold ourselves to priorities.

Achieving behavior enables us to progress toward our heartfelt goals, do something about the problems we see, and accept new challenges and appropriate risks. Like Alex, we want to put forth our personal best in service of something bigger than our own self-interest. We have a purpose, and we work to achieve it with excellence.

We can all increase our reliable and achieving thinking and behaviors through these steps.

1. Find Your Direction—
in Big and Small Things

We have seen that people can move in and out of three states of direction in life:

1. **I know where I'm going:** I know what my dreams, my goals, and my heart's desire are in this stage or part of my life. I know how I can make a positive difference. I'm living my dream, or at least know what it is and see a path to it.

2. **I'm going through a season of change:** I used to know where I was going (and maybe still do in some parts of my life), but something has changed and I'm trying to reestablish that sense of direction.

3. **I don't know where I'm going . . . but I want to:** I feel a bit lost or stuck, and I'm not sure what to do. The options seem overwhelming and I don't know how to decide.

Most of us have experienced the second and third states, and more than once. The great news is that change won't blow you so far off course that you can't find your way to a new and meaningful destination. And feeling lost never has to be a permanent state. If you're struggling with direction, remember that we're usually just a few decisions away from feeling confident about the path we're on and the difference we're making in the world.

Seasons of Change—Nothing to Worry About!

There are seasons in life, and we have found that whenever there's a new season of life for us, we experience the "confusion room"—a place full of anxiety because we can't see the answers. But if we can see a new season as a good thing, it can change everything for us. We tend to say that the "confusion room" heralds the "renewal room"—so embrace the confusion, find your way of being in that "room" with peace, not anxiety, and you will have more emotional and physical energy to move forward. Looking at a time of confusion as a new season transition takes a lot of the stress out of it, and we can cut ourselves some slack in not seeing the "answer." Know that it's okay if you can't figure out the answer *just yet*. Thinking in terms of seasons helps us feel that there will be an end point, too. What is *one thing* you can be *grateful* for as part of this time, right now? Be comfortable being uncomfortable. In our own lives (getting older!), we have found that nothing is ever lost when we move from one season to another: things are reshaped so we are prepared for the next part of our journey.

Dream-Building Inventory

Consider these questions our "dream-building inventory":

1. What's the biggest obstacle keeping you from going down the path you most want to follow? What's frustrating you?

2. If your path is truly blocked right now, what can you do in the meantime—with excellence and consistency—while you work on that obstacle? If the obstacle is one of money, what can you do to earn it or what can you sacrifice to save it?

3. What knowledge, skills, and wisdom do you need to seek out along the way?

4. What and whom are you going to put faith in to find support and peace during the journey?

5. What risks, stepping out in faith, will you need to take to pursue this path?

2. Trade Up to Achieving— from Controlling and Striving

One of the chief causes of frustration in everyday life is the rule or system that seems unnecessary, inefficient, or just plain dumb. The three forms you have to fill out, all with the same information. The four times you've been transferred from department to department to get an answer. Sometimes we take those rules and systems personally. We rail against them or try to work around them, sure that we know better. Sometimes that's true, and sometimes we're simply trapped BTL.

Gabrielle, a young manager for a retail chain, was pulled into that trap without realizing it. A few months before, her area manager had mapped out a new system to improve her store's performance. Gabrielle disagreed with the approach. She believed it was obstructing her from getting necessary things done in a particular way—*her* way. As her manager pushed her to put new practices into play, Gabrielle felt more and more controlled and offended. So she reacted by avoiding him and continuing to operate as she had been. *Why can't he see that this won't work?!* she kept lamenting.

Gabrielle's resistance and frustration were bleeding onto her team: her stress vibe was affecting everyone. Her young team members, many of them still in high school, were demotivated by her negativity—roster changes were demanded, calling in sick happened too often. Gabrielle had been proud of her achievements and her quick rise to a leadership position. Now, the stress was becoming overwhelming, she felt that everyone was fighting against her, and she was considering resigning from a job that she knew offered great opportunities for professional and personal growth.

Over drinks after work, Gabrielle was complaining again about her boss's demands. Her colleague took her aside and gave her an opportunity to see the situation in a different light. "Most of the stores in your area have not been performing well recently in a few key ways," her coworker explained, "and your area manager is responsible for changing that. He *has* to challenge you, Gabrielle—this plan is for *all of the stores*, not yours alone. I know you think your boss is a control freak," her colleague went on, "but have you considered that *you've* slipped BTL in the way you're behaving? Because your team is stressed out by your energy."

Momentarily taken aback, Gabrielle knew that what her colleague had said was true. She looked her colleague in the eye and thanked her for the genuine feedback. She decided to look at herself honestly, through the lens of S+T=B. "What is *really* happening for me?" she asked herself. "I'm BTL because I want to do things *my* way," she admitted. When faced with some truths about the situation, she was able to concede that her own plans to improve performance hadn't worked out. She had messed up, and while it had gone unsaid, she knew it, and her area manager knew it. Admitting this to herself allowed her to take responsibility and begin shifting ATL.

Gabrielle had a long history of being a high achiever—it was how she had risen to her position at such a young age. She had leveraged reliable and achieving behaviors to get where she was in her career. But, she realized, the situation of her store's slipping performance and her boss's efforts to correct it had triggered old templates that were subconsciously driving her: two years before, when she wasn't yet a manager herself, her store manager had implemented a new idea that didn't work, with significant consequences. Gabrielle realized how that had negatively affected her. For things to change, she had the opportunity to let go of the old templates, reclaim the gold within, trust her boss, and trade up back to achieving.

Swap Out Stress and Put Purpose in Its Place

Trading up can be difficult when we believe that we are *already* living ATL. We trust our intentions and can be blind to *how* we're trying to make those intentions a reality. Controlling can look a lot like reliable. Striving can look a lot like achieving, but both controlling and striving behaviors create unnecessary stress. Gabrielle was stuck in controlling and striving. She was attached to the outcomes of her store's performance as proof of her worth. When those fell short, which can happen to anyone, her self-worth took a hit and she was triggered. She became even more controlling and striving, somewhat desperate to prove that she could fix the problem all on her own. *Her* truth she perceived in her manager's efforts to guide her was, "He doesn't value my opinion or abilities."

When we're in striving and controlling, what we accomplish (and it can be quite a lot) becomes about us and what we need to prove, via comparison. When we are in reliable and achieving, what we accomplish is about what we can contribute with others and for others rather than to set ourselves apart from others. Instead of asking, "How does this elevate me?" we ask, "How does this move us closer to our shared purpose?" As Bob Goff, the author of *Love Does*, wrote, "We won't be distracted by comparison if we are captivated with purpose."

After the coaching from her colleague, Gabrielle applied a couple of methods to trade up from controlling and striving to achieving.

Stop, Breathe, Think, Behave: Recognizing the subtle signs in how she felt and the vibe she was putting out to others was a crucial first step in shifting her thinking and heart attitudes away from striving and into achieving. Gabrielle started by applying the *Stop, Breathe, Think, Behave* strategy so she could recognize the physical signs when she was being triggered and was resorting to striving coping strategies. She identified some specific moments when those striving strategies created tension and stress in meetings or for her team and intentionally practiced *Stop, Breathe, Think, Behave.*

What would my response and behavior look like if I were operating from an achieving thinking style and heart attitude? she asked. She would consciously relax her body and use less intense body language. She practiced shifting her tone of voice and the words she used—especially phrases like *should, have to,* and *must*—to avoid sounding demanding or blaming. Most important, she focused on channeling her natural high energy into a purposeful, positive intent that would help reduce stress for her and in her environment.

Build-Measure-Learn (the spirit of 80 percent): Gabrielle, now wanting to embrace and implement the new system, asked for help from her area manager. He suggested she adopt the Build-Measure-Learn approach of not having to have *everything* perfect before she got started. It was his way of releasing store managers from any striving behaviors and feeling they needed to be perfect when they implemented a new system. He knew it helped them trade up to achieving and remove the fear of failure. It was what he called the *spirit of 80 percent*. After coaching with her area manager, Gabrielle got her team together and said, "We'll implement the new system into our day-to-day operations, measure the outcome, and learn from the results. In the spirit of 80 percent, we might make a few mistakes in the first few weeks, but we will work toward getting it right. We can be a high-achieving team, so let's do it and have some fun!" Gabrielle went home that day feeling lighter, with a smile on her face, and so did her team! That night she realized how often in life she got trapped in striving, believing everything needed to be 100 percent perfect to prove her worth.

Before this day, she would keep pushing and pushing for perfection—stressing out herself and her team. Gabrielle realized that reliable and achieving behaviors helped shift her direction and get her back on track. Her manager was supportive and encouraging, and gave her the opportunity for even more development and responsibility. She made more time to coach members of her team. Her stress level dropped, her relationships at work improved, and her team was happier, more positive, and more productive, leading to improved results.

Striving—the "Evil Twin" of Achieving

In our modern world, it can seem that striving is rewarded. It can be how we prove to the world that we're "good enough." But this comes with a price and is also the path to incredible stress, burn-out, even depression. Mara calls striving *the evil twin of achieving.* Of course, nothing about it is "evil," but it does an excellent job of masquerading as achieving and then taking all of the joy and fulfill-ment out of accomplishment. When we're striving, we are attached to *how well* things are done, or *how good things look*, to support our self-worth and identity (think of the military-precision orga-nized linen cupboard with beautifully ironed sheets—and *towels*). This can even happen when we are making a salad, or worse, let-ting someone else make the salad after they asked to help us! We can suck the joy out of any experience for others, too, as they try to help but we continually tell them "you're not doing it right!" Really, let's face it: Is the entire lunch going to be a disaster because Johnny cut the tomatoes for the salad in a different way than you would?

Gabrielle realized that her templates led her to striving and per-fectionistic behavior. When we're in the grip of that, our stress energy impacts others—like Gabrielle's did with her team. A down-ward spiral starts, where *everyone* gets stressed and no one is en-joying the environment. If anything, people start to try to get away from it—like Gabrielle wanted to do, and some of her team tried to do by calling in sick.

Striving is one of the most insidious BTL behaviors, often made worse by our mixed cultural messages. Authors, psychologists, and management experts agree that striving for more can cause enough stress to kill us, or at least kill our joy in life, but what we see and hear around us is quite different. Do more. Be more. Have more. Get up at 5 a.m. Hustle. Be more creative. Post better pictures. Buy a bigger house. *Get the perfect life.* We tweet the latest wisdom about getting enough sleep—at 12:30 a.m. while catching up on the news of the day. We attend workshops on the latest tricks for im-proving our focus—with our devices set to silent yet still buzzing for attention. We're told to prioritize, to attend to the essential, to resist overcommitting, but one of the most common complaints we all share is not having enough time in the day to accomplish what we feel we "should."

Let's rebel against the glorification of busy-ness!

For some styles, especially striving and achieving, it helps to dig into the differences between BTL and ATL so that you can spot them more easily in your daily life. When we're stuck in striving, we do a very good job of justifying our behavior to ourselves and others, because we accomplish so much. So how can we spot striving behavior and trade up to achieving?

We would say that strivers may need a truth teller in their lives, like Gabrielle had, because it can be *so* difficult to distinguish between the two behavior styles, especially without some external perspective. But you can also be on alert for the common signs of striving and, using the strategies described here and in chapter 6, shift toward a corresponding behavior of achieving.

Finding Your Reliable and Achieving "Gold"

- When have you tapped into courageous humility and the gold in your character that enabled you to be reliable? What is a common thread in those situations?

- Where or when in your life have you been especially achieving? What drove you to go the extra mile to attain certain goals?

- What triggers or templates, voids or wounds are blocking you from being more reliable or achieving in certain situations?

Moving from striving to being high in achieving behavior isn't a guarantee that a person will also be high in reliable behaviors (although the two often go hand in hand). People who are high in achieving behaviors can also rationalize that those strengths make being reliable less important. The truth, though, is that confidence-building progress toward our goals requires both.

The Unreliable High Achiever

If you asked Justin how he wanted to treat people, he would tell you "with respect." He was friendly, easygoing, and genuinely well-liked. Unsurprisingly, he scored high in some of the love-driven behaviors. But if you asked people whether they felt *respected* by him, the responses were less positive.

When Justin got the results of his Indicator and saw some of the verbatim feedback comments, he was rocked. "Justin is often MIA." "It takes forever for him to get back to me on decisions." People scored him high on avoiding behaviors and low in reliable behaviors. They felt he was selfish with his time and didn't feel they could depend on him to help when they needed it. He could hardly believe it. It was the exact opposite of the values he held. And yet . . . he self-reported being poor in time management, in priority setting, and in consistent communication. He didn't see that being unreliable eroded his relationships and the positive impact he wanted to have on others.

Justin grew up with two parents who worked constantly in their business. Their family time was limited, and they rarely had much fun together. He held onto a truth that being driven or disciplined had to mean lousy relationships, little fun in life, and no time for family. At work, that template played out as resisting self-discipline and as a result, overlooking some simple strategies that could help him be consistent in meeting his commitments. He justified it by explaining that his job required him to be flexible.

Justin saw discipline as a punishment and feared it would make him rigid or boring. Maybe he could change his paradigm, we suggested, and look at it simply as a way to build consistency into his life, so that he could do what he said he'd do and finish what he started. Justin's journey was one of building consistent reliability.

Most of us have discipline in our lives already, even if we don't realize it—small systems for holding ourselves accountable. Most of us get up at about the same time Monday through Friday with an alarm. We use our technology to set up alerts and reminders or to keep our schedules straight. Sometimes, though, people need to crank up the discipline and structure in their day until consistency becomes a habit.

3. Be Ruthless with Time— Gracious with People

In thinking about S+T=B, Justin realized he was disciplined in *some* aspects of his life, and he had systems in place to help him be reliable in some key ways. For instance, he was rarely late for meetings or calls, because he relied on his calendar. But he did have a tendency to go on too long in meetings if he didn't have another commitment after, disrupting other people's time. He worked through finding the situations that triggered him into BTL, and he identified a few of his templates and truths. Discovering what was happening for him, he built more structure into his days so that he could be proactive rather than reactive.

Calendar reminders and e-mail management: Justin realized that e-mail was a big issue. Instead of seeing it as a helpful assistant, he often perceived it as an angry taskmaster. To be more reliable, he blocked out thirty to forty-five minutes in his calendar in the morning, middle, and end of the day, just to deal with e-mail. As a result, Justin procrastinated less on decisions and on communicating difficult information. The time he spent dealing with e-mail actually shrank. He also used his calendar reminder system to keep him accountable and help him keep on top of important things to remember.

Be ruthless with time and gracious with people: Too often, we let the pressure of time drive behaviors that aren't effective and get the better of us. Being ruthless (in an ATL way, of course!) in how we use our time, consistently prioritizing to stay focused on our purpose, can help us strengthen our reliable and achieving thinking and behaviors. But if we are not also gracious with people as we do this, we will either end up BTL, or at least *seem* BTL. When there is a time clock going off in our head, having the composure to not rush and be calm is a wonderful character strength.

Justin planned his availability in a consistent way by *frontloading his intentions* in his interactions with people. If you don't have the time or focus to be helpful in that moment, it's okay to share that. It's more than okay, it's a way of showing your respect. "I know this is important," you might say, "and I only have five minutes right now. I want to give it the attention it deserves, so how about we meet [give an alternative time]. I'd like to talk about it more then." You're thus communicating your respect for the other person's needs and time.

Over time, these new systems helped Justin build new habits. Those habits shifted his thinking and his heart attitude. The more reliable he became, the less he resorted to avoiding behaviors. With some effort, people came to see his behavior as dependable and respecting other people's time, which is what his heart had desired all along.

The key is to look at the systems that already exist in your life, figure out where you can adopt a reliability habit boost, and merge the two.

Nothing can be more satisfying than naming the purpose in your life and pursuing it in an ATL way—operating in your personal best, your PB. That satisfaction is deepened when we find the joy that comes as we connect with others on a genuine level. That's what lies ahead in the following chapter.

CHAPTER 9

Connecting with Others—
Relating and Encouraging

I believe in businesses where you engage in creative thinking,
and where you form some of your deepest relationships. If it isn't
about the production of the human spirit, we are in big trouble.

—*Anita Roddick, founder of The Body Shop*

We can almost guarantee that right now, you have a relationship you wish were stronger, deeper, more open, more loving. It is the essence of our being, to want to feel connected to others. We were created for relationship. We're not supposed to do it alone. We are tribal by nature, yet in our "selfie" world, we are becoming more individualistic, uncommunicative, buried in our mobile devices, and more and more critical and judgmental of others to safeguard our own self-worth. When we shift to a "we" focused world we experience the joy of connection that resonates with the gold and love in our heart—and engages others. It gives us a deeper sense of meaning.

Anytime Olivia walked into a room, all heads would turn. This charismatic, skilled legal counsel was beautiful and stylish, at the top of her game. Her sharp sense of humor reflected her intelligence— and her boundaries. That sarcastic wit could be scary at times. So much so that when Olivia's colleagues completed an Indicator on her, there wasn't a whole lot of ink showing in the love quadrant results. In fact, they rated her less than ten percent in the love behaviors.

Olivia was utterly floored. "In a nutshell, these people who mattered to me did not see me as relatable, encouraging, or compassionate. What they did see was off-the-chart levels of sarcasm and controlling behavior. And I didn't see *anything* of this. I thought I was being friendly and funny—after all, I've had people laughing so much, we have high-fives, all joining in on the fun."

At first, she was pretty steamed up. "Who are these people? How dare they say this about me? I'm not that person!" It took a very long phone call with her husband, himself an eminent lawyer then on an international work trip, for Olivia to come to the conclusion that maybe the results weren't wrong after all. Ninety minutes later, she was ready to concede that perhaps there was some truth in the results and there were some development opportunities.

The wonderful AND at the heart of Olivia, though, was she had a large circle of friends and a tight-knit family. To see the feedback from her colleagues shocked her to the core. That was not who she saw herself as. At first, she simply couldn't understand why people at work were seeing her in that particular way.

Once she had absorbed the initial shock, Olivia talked with us about what might have brought her to this point in her life. She had been set on resisting any deep reflection. She felt she already knew herself pretty well, and the truths she had always lived by served her just fine. Olivia had been raised in an upper-echelon family populated by doctors and corporate lawyers. As she put it, "I come from a long line of people who do not easily share their feelings or—God forbid—exhibit vulnerability. That is *not* what we do. We all love one another in *our* family's way of doing that—through talking about politics, global events, sports, and just about anything that can be debated. We love a good laugh and a good argument. We have great times at Thanksgiving, we all get along well with each other's spouses, and the cousins play well together. We're all friendly. I can't believe that my team do not see me in the same way."

By choosing a career in law, Olivia naturally gravitated toward a very combative occupation in which displays of emotion were not encouraged. "As a young lawyer, one of the first lessons you learn is to protect yourself: in debate, in negotiation, with opposing counsel, and even among your own team," Olivia explained. "Later as a corporate lawyer, you make every effort to avoid allowing emotion to cloud your decision making, particularly where litigation is involved. Showing emotion is *not* the role of a senior leader. Especially legal. That's for the sales and marketing team," she said in her sarcastic way. Because of her upbringing and her professional training, Olivia believed her primary responsibility was to avoid being vulnerable and emotional at *all* costs.

So profoundly had Olivia closed herself off, she summed up her outlook on life like this: "Let's face it, if you've been around a while, you learn to protect yourself." That was her theme for life, but it was a *my* truth versus *the* truth belief. In her heart, Olivia had the opportunity to understand the power of the AND, to shift *from* "Let's face it, I've been around a while—I've learned to protect myself" *to* "I've been around a while—I've learned being vulnerable is okay at times."

As she started to unravel the results, Olivia came to understand a lot of it was her own doing. She didn't prioritize connecting with people or actively establishing relationships at work. The only relationships she had there were transactional, and in her eyes, this meant she was good at her job. Her priority was being logical and very critically evaluative so she could protect the company. Along the way, though, this tactic reduced her relating, compassionate, encouraging, and developing behaviors and increased her striving, controlling, and sarcastic behaviors. Hence that close-to-zero love quadrant result from her peers and direct reports. They thought Olivia was very good in her functional expertise, but they just couldn't connect with her. In fact, some of them were outright scared of her and thus did not approach her for help, advice, or opinions.

Olivia made a decision to take the feedback as a gift and to act on it. The way she saw it, she wasn't going to change who she was, but she *was* going to address those behaviors that were clearly preventing her colleagues from seeing what was in her heart and her best intentions. She knew she could learn and practice being more relating and encouraging, even if it was going to feel awkward. Her first step was to switch the paradigm she had constructed, which went something like this: "It's a waste of my time going around talking to people and getting to know them." She decided to change it to: "I *do* enjoy getting to know people." Being a highly disciplined person, Olivia knew she had to take a structured approach to changing that paradigm. In her calendar she blocked out fifteen minutes, three times a week, just to walk around and say hello to a few people. She knew that eventually, if she kept it as an authentic desire, doing it often enough and consistently enough, it would become part of her, and that getting to know people would build trust and relationships.

The first few times she tried it out, she found it *challenging*. In the first week, she found that even though she was *trying* to be relating by asking questions about people's lives, she wasn't really all *that* interested in what the other person was saying. Her focus kept wandering. "Right," she thought, "next time I need to forget about my timelines, and just allow myself to really concentrate on what this person is saying. Find something that's important to them in what they're saying to me, and focus on drawing them out with questions." So she concentrated on focusing her attention more on each person—and found that she became more genuinely engaged in people's conversations.

A few weeks went by. Olivia was feeling more confident and people were engaging with her more. Then she had a low week, where she found herself back to not focusing her attention again. This time, Olivia knew to ask herself what was happening for her. She realized that she was in the middle of a tough legal matter and that her "time bomb" was back in her head—putting task right in the forefront of her mind. "I was focused on my own work deadlines as I was listening to people and trying to control those thoughts about having no time. This was distracting me from focusing on the person," she explained. So she just kept practicing, and week by week it became easier for her. To this day, it's still something Olivia works on, but she knows that just going through the process is good for her. She has created deeper connections with other people. In and of itself, that helped draw Olivia more and more out of herself and into ATL thinking and behaving. Ultimately, she has a more joyful experience in the workplace (and so do her colleagues, presumably!).

Olivia's story shows us a person who, over time, had constructed a set of walls around herself that prevented her from really connecting with her team. Without a shadow of a doubt, she is now on a journey, and she is still working on being more comfortable with genuine relating rather than relying on sarcastic humor to try to connect with others. Her fifteen-minute walks have become a habit she enjoys, and she concedes, "I may have moved the needle on sarcasm some!"

Olivia in her private life helps the local community with some pro bono legal work. Soon after this experience she had the revelation that the word "community" is two words, common + unity, and she started to see how that applies in all relationships and in all environments. She grasped the truth that difference equals difference—it doesn't equal wrong. Relationships in life and work are not simply based on people who are just like us. We often need to build relationships with people who are different from us, very different. So, Olivia took her community heart and personal values to the workplace, and engaged even further with her team, loving them as she does the community.

Recap—Relating and Encouraging

These two styles are next to each other in the Heartstyles model because they work together to encourage the inner character and heart of people. As Stephen Covey explained in his book *The 7 Habits of Highly Effective People*, it's about making deposits into another person's emotional bank account, their character.

Most people want to be known and be heard—to know they matter. In the work context, we particularly want to be known by our leader. We want to know that getting up every morning and going about our work each day is all worth something more than just a paycheck. We want to know our leader cares about us and feels we are significant. From a leading/managing perspective, relating is also about letting people know how much they matter and are valued through taking a genuine interest in them as people, not just their functional expertise.

This is why, as people grow in their management and leadership responsibilities, shifting from task to people, they change their paradigm about what is a successful use of their time. Because when you're stuck in "task mode," you think the only way you can be successful is by ticking off your to-do list of tasks; therefore you think you have to be focused on your to-dos, not on other people. This is the point Olivia had reached: early in her career it had all been about task, task, task, which made her transactional, transactional, transactional. She had no time for relational. It just wasn't a priority for her because she was too laser-focused on getting things done. Intellectually, she understood relationships matter, too, but behaviorally she wasn't delivering on it, and she needed insight for transformation.

Although Olivia found she needed to make a conscious effort and put structures in place to shift her focus to relating, it was a classic case of getting in touch with her community heart to live it out in the work context. Olivia was committed to changing her behavior through doing the behavior, and over time it became more authentic.

You'll know from your own PDG that the relating style measures the extent to which you are focused on getting to know people at a real and sometimes deeper level and valuing others for who they are. People know you for your solid interpersonal and social skills and your effective communication, and they feel you really care.

The encouraging style measures your genuine interest in helping others grow within, to grow their character and their inner worth, knowing they are valued. You do this by being comfortable giving supportive praise. Showing confidence in others (and in yourself), you can effectively communicate your belief in them and their ability—through both verbal and nonverbal means.

You may tend to be a good listener and encourager—listening more than talking. Others will see you as someone who genuinely compliments others on what they have done/said. People may also know you as someone who is honest in their compliments, and they appreciate that.

A CEO once said to us, "If I get friendly with my staff and compliment them, they'll want more money." Some people in business won't encourage, because they think, *They might take my job. They'll get a big ego, and that's no help to anyone.*

Every great organization we have worked with has the power of recognition and encouragement built into their culture. It is simply one of the most powerful ways of motivating people, building people, and retaining talent. It has the power to help build the inner person. It's a humbling experience because you realize it's not about you—it's about you encouraging someone else. However, we do need to check our own sincerity. Signals that help you know it's about you are when you find yourself not really listening and thinking of other things; when you forget where you're up to in a conversation; when you're easily distracted; when you swing conversations back to you; or you interrupt others or take the conversation off on a tangent.

Part of effective encouraging behavior is monitoring what we say and how we say it. The most effective tone of voice for encouragement is very straight and non-fluffy. Terms such as "You're awesome!" offered in a high-pitched tone can come across like approval seeking (and at times, if we're honest, it may well be!). Choosing the appropriate words and voice tone for each person we are encouraging is also important. How do they like to be encouraged?

Having the courage to build courage in others through genuine support and praise—this is what true encouragement is all about. Expanding and building character through "in-courage-ment" comes from the fact that we are all meant to put courage into other people's character. You can also practice encouraging up—to your parents, your boss, your coach. The best of people, their potential, their release from fears, their unlocking of respect, loyalty, and honor will come about through encouraging others. Encouraging doesn't have to take a lot of planning: it is something you can do spontaneously. It can be verbal or written; you can do it over WhatsApp; you can do it face-to-face. You can grab a Post-it note, write an encouraging message on it, and go slap it on someone's desk. And the power of the handwritten thank you note sent to colleagues, suppliers, clients, or friends after an event, as old-fashioned as it is, is still one of the most sincere ways of relationship building. For many, it's more meaningful than an e-mail or text, because it happens so rarely—a case where analog can trump digital!

Approval Seeking or Relating— What's the Difference?

A while back, we interviewed an experienced trainer to help grow our workshop facilitation team. Stephen was impressed with him at first because he came into the interview having many little-known facts about our organization and Stephen's background. He had done his research. He communicated with enthusiasm and was impressive in his experience. Mara's perspective was rather different, though. He had done the same deep research on her, but she also noted his body language and how he brought every topic back to himself to seek her approval. He positioned himself to be within her line of focus no matter where she turned, to the point of it feeling uncomfortable. We know his good intention was to impress us, but the overly nice approach became too much and what started out looking like relating ended up coming across as approval seeking. That just would not work with us or our clients.

The word "approval" has the word "prove" buried in it. "A-prove-all" is closely associated with competitive, but much more subtle. "Nicer," yet all the while endeavoring to prove how good one is—seeking validation. It's easy to confuse relating with approval seeking in our own behavior and in others, which is why we have to explore our heart attitudes and thinking styles, especially when it seems our behavior isn't actually helping us connect with others.

Find Your Relating and Encouraging "Gold"

1. In what examples in your past, in or out of the workplace, have you found love and respect and "gold" in your character to be relating with and encouraging others you get along with well?

2. In what situations in your life could you practice being more relating and encouraging with people who are more difficult to connect with?

3. What do you believe could be the triggers, templates, or truths or voids, wounds, or vows blocking you from being relating and encouraging?

Approval Seeking Is the "Evil Twin" of Relating!

When We're Approval Seeking	When We're Relating
We do things for the approval of others, and get disappointed when others don't show an expected level of appreciation.	We get along well with others and are friendly and approachable, without being needy.
We are a people-pleaser.	We are a people-connecter.
We can't say no because our need to be approved of by others is greater than our ability to set appropriate boundaries.	We are able to set appropriate boundaries with others and explain why.
We connect with others with the hidden intent of seeking their validation and affirmation—we need to be liked by everyone.	We connect with people by being interested in other people's information and interesting to others in conversation. We like to be liked and like others, but we don't *need* to be liked.
We are too fearful to disagree with a point of view; thus we end up agreeing with everything, or saying nothing so we don't get disapproval.	We are prepared to politely disagree with a point of view.
We spend an inordinate amount of time feeling we are being judged, so we spend lots of time thinking about what others think of *us* and rehearsing conversations in our head prior to interacting with others.	We have genuine confidence within, so we are not concerned about what others (may) think about us. We realize others may not even think about us, and that's fine!
We want to be associated with other people who are famous/interesting/important—even vicariously associated with them to "bask in the reflected glory"—we *need* admiration to build our self-worth and self-esteem.	We accept people for who they are as a person, not because of who they are socially or what they do professionally.
We behave like those around us to fit in and be accepted. We have over-emphasis in our voice tone, adjectives, and body language without the self-awareness that our behavior is extreme—we *need* attention and praise.	We behave in ways congruent to our personal values and we appropriately respect other people's values.

Encouraging on the Run

How do you find time to spend with people? For many high achievers, relating and encouraging can be the lowest ATL styles, for that exact reason. Time! It's almost like a time bomb going off in their head—because they've always got something on the go. So how do we balance spending time with people with getting things done? If only we had the time, we say!

A CEO we know well is the master of this—balancing people and task. He is the highest-ranked person in a massive multinational company, and yet he will get out of his office for an hour most days to walk around and just chat with people. He will ask people about their lives—how's the family, where have you been, or where are you going on your next holiday? I hear you did well with that presentation last week. Tell me about that.

Knowing him well, we think one of the reasons he is so good at this is that he doesn't spend *hours* with people—he has little conversations with people, even for a few minutes, and because he is genuinely interested it makes that person, or those people, feel they *matter*.

To continue being ruthless with time and gracious with people, here's how: start by looking at your calendar. Look at the meetings that go for an hour. Could any of them be reasonably cut to thirty minutes? If so, you're creating thirty minutes of time.

The other question is, am I really meant to be in this meeting? You can start to delegate. Even if you still attend the meetings you feel you have to be in, consider how you can chop them down by ten to twenty minutes, so you're making up some time that you can then take and use for walking around. When we really analyze our meeting process, we often realize we could be a little more efficient in our time usage, without it robbing us of our encouraging time!

Three times a week, give yourself ten minutes to go and target certain people in your environment. Say to yourself, "Ten minutes Monday, Wednesday, Friday," then put it into your calendar so you have to do it. You start (appropriately!) shaving a little time off your meetings, and then use that "extra" regained time to perhaps stop by someone's desk, or at the coffee station, to have a two-minute "Hi, what's happening for you today?" or "Thanks for taking the lead on that project" chat.

Truths and Tools for Relating

Let's face it, until any of us sees the benefit of being ATL is greater than the benefit we currently think we're experiencing by living BTL, nothing will shift. In theory, the practice of focusing on another person will always work, but you still need to feel it's a beneficial thing to be doing.

We can easily think of relating as a behavior that either comes naturally to us or is entirely alien to the way we interact with the world. Not so! Here are five tips to turn the dial for people who want to improve the way they relate with others.

1. "There You Are" versus "Here I Am"

When someone walks into your office and they want something, you can either think, "Here *I am*; what are you doing annoying me right now, because I haven't got time for this." Or you can change the paradigm to "Oh, *there you are*. How can I be *of service to you*?" Value the other person as significant and your genuine relating energy will be well received. Likewise, when you go to a meeting, turn up at a party, catch up with friends, or arrive at the sports game: Is your energy all-about-you or all-about-others? Is it *there you are* or *here I am*?

Anytime you direct your attention to others, what you're really doing is shifting your heart to humility and love, being fully present and changing the atmosphere around you.

- Make sincere eye contact with people when talking to them. In your heart, love, respect, and honor the person—you will be less likely to get distracted.

- Active listening allows people to finish what they are saying, without you interrupting or trying to complete their sentence. Ask questions to ensure you understand. Take notes.

- Try to remember what people have discussed with you so you can ask them about it the next time you see them.

- Warning! Be mindful to not bring every conversational topic back to *you*!

2. The I, the Us, and the We

All of us, in our hectic, sometimes overwhelming lives, may stop devoting time to the important relationships with our spouses or partners. This can be especially true when we have kids. Everything becomes about work, the house, homework. We lose touch—with each other and with ourselves. The I, the Us, and the We concept brings our focus back to the three areas of our relating lives:

I = **my personal growth journey**—the things I do, like, and need for me; the things I value as a person; my personal space needs. What I need to reenergize.

Us = **our relationship**—as a couple. What made us fall in love in the first place; it's only about the two of us and what is happening in each of our hearts, our struggles and joys of the heart to share with each other. The intimacy of being vulnerable with each other.

We = **who we are together**—*everything* that goes on outside of us as a couple. Our children, family, friends, work, sports, interests, hobbies.

Every week, make time that is strictly **I-time**—getting some personal space, spending it in whatever way allows you to come away feeling a little recharged, refreshed, and reconnected with who you are, your heart and purpose.

For I-time, consider these things:

- Work with your partner to give them their I-time and for them to do likewise for yourself.

- Think about what brings each of you joy and energy in that time.

- Figure out how much time you both need. This sometimes needs to adapt to changing circumstances. Before children and a busy work travel schedule, some people would play golf every Saturday for their I-time. But now with a young family who have hardly seen you all week, sometimes you have to find an alternative.

Every week, make time that is strictly "us-time"—being fully present with your partner or spouse. This is all about "intimacy"— "into-me-see." Spend time in intimate conversation with real heart exchanges.

Look at improving us-time:

- Date night as often as you can. That can be going out or staying in and creating a lovely atmosphere at home—even just getting all the junk off the dining table and creating a lovely atmosphere. A couple of candles and takeout make a big difference!

- Think about what us-time looks like for you as a couple. How often can you make that time? Whatever works for your situation, agree on it, make it a priority, and stick to it!

- Listen with your heart. If you're going through a tough time about something (at work, with the kids, family, or friends), then being able to talk it through with your partner is us-time. When together, use phrases like "What's happening for you?"

"What's *really* happening for you?" "What's happening in your heart?" "What can I do for *you*?" "What do you need from me for our us?"

- Watch out for flipping conversations into problem-solving mode. That doesn't mean *not* talking about it, but it does mean keeping on the level of speaking about and listening to, each other's heart and emotions, thoughts, and concerns. It's just keeping a boundary, to not let solution-focused conversations sabotage the "us" intimacy time.

- You may need to agree as a couple on one night in the week you stay at work late. Expectations are clear, and other nights you can be home fully present.

Almost every day we are consumed with **we-time**. This is time spent doing life: work, the family, friends, social engagements, sports, having fun. Most of life gets eaten up by we-time and we lose the us and the I.

Here are some thoughts on improving we-time:

- Understand that we-time is what you do *with* others that's *about* others—not about your relationship. The obvious example is being parents to your kids.

- Make we-time special but not so totally consuming that it leaves no time for I-time and us-time.

- Make we-time an opportunity to do something for others in the community together with your partner, a good friend, or family member.

3. The Mighty Four Words for Connecting

We call our approach to engagement the Mighty Four Words. The next time you really want to show you care, you're interested in knowing more, and you want to understand somebody's situation at a deeper level, ask, "*What's happening for you?*" This question is not asking what do you *think* or what do you *feel*; it's neutral. "*What's* happening *for you?*" It kicks off a different kind of conversation, one that encourages the other person to honestly share. Usually people ask "How are you?" or "How's things?" and the reply is "Fine, awesome, great." We hear these questions too much, and our responses are preprogrammed rather than authentic or vulnerable. It doesn't create real connection.

Sometimes the response can be "I don't know"—that's usually a habitual neocortex response that's either a conscious blocker to stop them from saying what they're really going through, or a subconscious blocker to protect them from pain or discomfort, or a simple lack of confidence. So, you can then say (casually!), "What if you *did* know?" We can pretty much guarantee this simple question will unlock the subconscious block for the person. Well, unless they are a teenager . . . try this with a teen and you might get a snappy response, "I just *told* you I *don't know*!" as they stalk off. For most adults, "What if you *did* know?" is an amazing unlocking question, as it goes into the limbic, where our emotions and memories are. It can initiate a deep and meaningful conversation.

4. "Frontload" for Effective Relating

Spontaneity is a beautiful thing, but sometimes relating does have to be planned. It has to be intentional. This isn't about faking it, though: in order to strike a balance between connecting in the workplace and getting the job done, you need to plan your time.

When you're in a meeting situation, sometimes you've got limited time to get through the agenda. Everyone secretly is on high alert because they all know the agenda will not get the full run through. The first part of the meeting could be eaten up with social niceties, time will run out, then the pressure of lack of time can drive us BTL, and someone is going to leave frustrated.

Here's an opportunity to use the frontloading method we mentioned in chapter 8. You might say, "Hi everyone, we all know we're on a tight deadline today. So if it's okay with you, how about we just get on with our agenda to save time?" Bang, off we go! Everyone understands the spirit of the meeting at the beginning and can get on with the agenda. That doesn't mean we take the interpersonal connection out of the meeting, but it gets us on to task quickly and diminishes the perception of controlling. Frontloading takes away the controlling.

We as people all want connection. Our high-tech world allows us to keep wonderfully connected with others through the Internet and social media channels. Distance is no longer an obstruction to relationships. Sometimes, though, that connection is being *disconnected* by our mobile devices—excuse all the puns! The statistics of how much screen time we spend are staggering. We need to check ourselves sometimes: Are we too busy attending to our mobile devices to give our full attention to our colleagues, friends, family, or children? "Technoference" is a word for those minor everyday intrusions or interruptions that technology devices bring to our interactions. You see it in restaurants: a group of friends around a big table, each transfixed by their own screen, or, sadly, couples out on a date. And you see it in office spaces: one colleague talking to another whose head is bowed over their phone. Some practices to manage "technoference" we have gotten positive feedback on include:

- Set expectations about taking any phone calls or e-mails during meetings and in the evening when you get home.

- If you know you have an important call coming in, frontload it when you check in on the meeting.

- Switch devices onto airplane mode—during meetings, at home, and especially on date night—for at least forty-five minutes while you eat and connect.

- Leave your phone in another room for the first hour you're home so you can be present.

- Learn how to program the e-mails you write in the evening to send first thing in the morning, so you're not sucked into a back-and-forth with colleagues at all hours.

Mobile devices can allow us to express our deep desire for connection. We do, however, need to be mindful of them not taking our attention away from the very thing we desire—connection with others in a real and meaningful way. When those mobile devices become "the devil in your pocket," they become destructive to relationships and rob us from giving people our full attention, or rob us from another's attention to us.

5. Listen to the WIFE!

So many people say to us, "I don't know how to do small talk. I want to show interest, but I don't know where to start." Let's look at a simple process to help develop some basic go-to questions you can have up your sleeve in social situations that will help with relating and connecting with others.

We've brought that down to an acronym. And the acronym is WIFE. Listen to the wife.

"W" stands for work. One of the easiest places to start to take interest in others is asking them about their work. Think about asking: Where do you work, what's it like working there, what were you doing before? A warning, however: depending on your profession, you might need to be careful that you don't come across as self-serving. For example, if we asked people about the leadership and culture of their workplace, it could look like we're pitching for business. If you sold software or provided recruitment services, you would need to choose your questions carefully. But at least start with a work question, as it is familiar to people and usually easy to talk about.

"I" is for interests. What interests do you have, hobbies, vacations, what do you do in your spare time, what would you do if you had more time, what is your dream vacation destination, anything planned for the weekend? Who/what inspires you?

"F" is family and friends. Do you have a partner? Children/ ages/what do they do? Where were you born? Where did you grow up, go to school, college? What is that country like? Extended family, where are they from? Do you have any brothers/sisters? What do your friends do, do you get together much, do you have friends/ family close by/all over the world?

"E" is entertainment. What do you do for entertainment? Movies—which have you seen lately? Are you a sports fan? Do you go to the theater? What's your latest book? What is your favorite movie/book and why?

Last, if you're going to be coaching someone and help develop them to be their best self, you will be way more effective when you have built a trusting, safe relationship through relating with and encouraging them. In our often hyped-up and judge-y world, where social media tells us exactly how our lives *should* look, sometimes all it takes to connect with others is simply not judging. Respect and acceptance of others are the heart of relating and encouraging. Once we set aside our thirst for more "likes," that's when we open the door to true connection. When we give out of the abundance of our heart, our gold, our love and respect, others sense the sincere honor we project that *most of the time* doesn't need *a lot of time.*

CHAPTER 10

Growing with Others—
Developing and Compassionate

A coach is someone who can give correction
without causing resentment.

—John Wooden, legendary US basketball coach

Every one of us has the desire to be an effective parent, partner, leader, teammate. And we *are* all those things, when we tap into our compassionate heart attitude—looking beyond the behavior to understand the heart and motivation of the person, caring enough to offer constructive coaching that helps people reach their potential.

In 2004, one of the most devastating natural disasters struck in the Indian Ocean. The massive earthquake that led to a tsunami destroyed coastal communities throughout Southeast Asia and cut short at least 230,000 lives.

Living in Sydney at the time, for us the tragedy was close to home, even more so because our twenty-one-year-old daughter, Tamara, was traveling the beaches of Thailand with two Aussie friends. We were frantic for twenty-four hours, not knowing her whereabouts, and then Stephen received a five-second call from a strange number: "Dad, I'm alive." Thankfully, she had been traveling with two experienced surfers. That morning her friends had gone out early to get breakfast, and walking along the beach they saw the water receding *away* from the land. They knew something was terribly wrong. They ran back to the villa, grabbed Tamara, and ran to safety—there is another book in the continuation of that story.

Soon after, Mara, who has a PhD in clinical neuropsychology and lots of professional experience in trauma counseling with a wide range of people, received a call. She was asked to come to Sri Lanka to help counsel children and adults who had survived and those who were there to help. The night before she departed, we went to dinner for us-time, a bit of a tradition when we're going to be apart for some time.

We sat down, ordered wine, and began to talk about her trip. Eventually, she paused and said, "You know, darling, I'm really glad I'm going to help. But I feel like you may be a bit jealous." Why did she feel that way? Maybe it was the edge to Stephen's tone when he said, "I think it's great *you've* been selected to go help these people." Maybe before the dinner he'd pulled out one of his standbys when feeling threatened by Mara's credentials: "Well, *you're* the psychologist." But at dinner, what did he say? "No, no, no. I'm not jealous—I'm happy for you." He didn't realize he was in the grip of an AND moment—he *was* genuinely pleased for Mara, *and* . . . jealous.

One of Mara's character strengths is that she doesn't have to prove she's right, so we moved on and had a lovely evening. One of Stephen's character strengths is that he genuinely listens when people give him thoughtful feedback. He spent the twenty-four hours after Mara's flight pondering what she had said. And what was his revelation? "Of course, she was right. I was jealous," Stephen admits. A few years before, we had been involved in trauma counseling a group after we came across a murder-suicide in the middle of the Australian bush (there's another book in *that* story as well). "Having worked alongside Mara in trauma situations, some of her processes, especially those focused on healing the wounds of the heart, were developed by both of us over the years. 'Why didn't they ask me, too?' was the question in my heart."

Mara did everything right that evening. She wasn't focused on "who's right," she was focused on "what's right." She found the perfect moment of truth (MOT)—not while she was packing or getting in a cab, but while they were relaxed and focused on each other. She wasn't controlling but approached the topic with compassion for Stephen and with a genuine desire to help him develop his character by planting the seed. Mara behaved out of love, and Stephen knew she had his best interests at heart. Because of her behavior, he felt the safety and the freedom to recognize the truth and find the humility to apologize. She made a real, positive difference in their relationship by helping Stephen take an opportunity to grow his character.

In our work we've seen bosses, coaches, and parents help other people to achieve transformation through developing and compassionate heart attitudes—like the father we met in chapter 4 who coached his daughter from a crippling fear of math to becoming an A student, or the leader who coached a team member to leave the organization with dignity, self-esteem, and confidence.

Let's admit it. Coaching someone or giving someone feedback for improvement to develop them is one of the hardest things to do in life. We are fearful of offending someone, crushing their confidence, or making things worse for them. We're also concerned about their reaction of potential anger, tears, or even legal action. The irony is that most people do want to be coached; they want someone to believe in them and walk with them in their development.

Recap: Developing and Compassionate— "Character Coaching"

These two styles are next to each other in the Heartstyles model because the most effective way to develop others is with the heart of compassion, to meet people where they are at so they can grow.

Developing has a lot in common with a growth mindset, the term developed by Carol Dweck, professor of psychology at Stanford University. Developing behavior encourages others to believe that intellectual abilities, social attributes, and character within can be grown. The greatest sportsperson still has a coach. They have a growth mindset. This is what we call "character coaching." It takes character to coach someone and it takes character to receive coaching. Instead of thinking of it as confrontation, think of it as *care-frontation*. If you care for someone enough, love them and believe in them, and want to help them be the best version of themselves, how could you not want to help develop them, as Mara did for Stephen?

Compassionate is loving objectively. Compassionate *does not* condone ineffective or negative behavior—it doesn't mean agreeing with it and saying it's okay or right. The real heart of compassionate is observing people's behavior, not judging it, then looking past it into the heart, and asking yourself, "What is *happening* for that person? What *happened* to that person?" You're trying to look at the *whole* person, and as you look at them also understand that they are an AND. It's being able to say, "I believe people are fundamentally good, have good intentions, and their BTL behavior is not who they really are—it's a coping strategy to protect themselves from *something*. So *what* is it?" It takes strength of character to have the strength of compassion. The less developed our character is, by nature the more judgmental we become.

If you want to love objectively, it helps to think like a psychologist. One of the things psychologists are trained to do is to use their listening skills with developing and compassionate as their filter. Let's say you're in a counseling situation. Your client comes in and tells you all about a particular state of affairs. In this first conversation you're trying to get a sense of who they are, but you're also listening out for clues that help you build a bigger picture. As much as anything, you're listening for what they *don't* tell you. The gaps in things. As a psychologist, Mara likes to think of it as a puzzle. In her head she sees a set of puzzle pieces with notches in them, and she needs to slot them together to build the picture.

You're sitting there listening objectively. Rather than just empathizing with their problem, you're trying to figure out what other pieces are contributing to their perception of their issue, and therefore what information you can gather to build a picture of this person as a whole. It's about trying to uncover some objective truths so you understand how that exists within *their* (subjective) truths, and seeing where the gaps are so you can help them, with compassion and care, to bridge those gaps.

The word "compassion" comes from the original Latin *com pati*, meaning "to suffer with." Among emotion researchers, it is defined as the feeling that arises when you are confronted with another's suffering—and feel motivated to relieve that suffering. It's that second part that elevates compassion beyond just a warm, kind feeling: that drive to do something to help, to take action.

When you go to where love is, you find compassion. We tend to go to where love *isn't*, and that's judgmentalism. All of us can be so judgmental at times. If I make you *wrong*, then I must be *right*. And that feels so good! We tend to judge the world, and that gives us a sense of security and a sense of authority and feeds the ego need—but once again, it's the counterfeit at work! There's a difference between judgmentalism, which is BTL, and discernment, which is ATL.

Tips for Compassionate and Developing

As we go through the tips on coaching and developing people, remember there are two keys to effectiveness—sincerity and wisdom. If you use any process or method without sincerity and are not wise in where and when you coach someone, it will simply not work. When you are sincere *and* wise, people will trust you as an effective and caring coach who is direct and challenging yet supportive. They will know you have other people's best interests at heart in wanting to help them develop professionally and personally. Getting to this point of trust takes time. You may have to put a lot of "deposits in the emotional bank account," as we described it in chapter 9, by relating to and encouraging others.

Developing people is done in so many different contexts, and there is not a one-size-fits-all process—from small, day-to-day, on-the-run "drink soup while it's hot" feedback and coaching in the moment, to monthly catch-up sessions, to "big" coaching moments. There are no hard and fast rules, though. It's totally up to you to dive into these ways of thinking as you need them.

Here are five tips to help you master character coaching.

1. It's All in the Timing—Moment of Truth (MOT)

Have you ever tried to break open a green, unripe walnut? It's almost impossible. Its tough outer shell won't budge. You can use a hammer, which might bounce off or just squash it. But how easy is it to open a nut that's ripe, with just your hand, to get to the heart, where the goodness is? You cannot develop people, even with compassion, until you have found a moment of truth (MOT), when their heart is ripe and open to what you have to say. This is what you saw in our opening story about the two of us wrestling with a little professional jealousy. Sometimes you have to build to an MOT, creating a meaningful connection with relating and encouraging behaviors. But even once you've done that, finding the best MOT takes thought.

It's also about looking for the opportunity, having courage, and then creating that MOT with someone. This takes self-awareness, awareness of others, looking beyond the behavior, and seeking to understand what is happening for the other person to be causing the current situation—applying a potent blend of S+T=B and compassion for them.

One of the surest ways to get to an MOT is through an ETA (because too many three-letter acronyms are never enough!). In this case, ETA refers to environment, timing, and approach. It can refer to the wisdom of when to bring the feedback topic up, and also when to actually have the discussion.

Environment

- Think about what sort of physical environment will suit the discussion.

- What does the person need for them to be comfortable and safe? It could be done privately, on-the-job but not distracted, in a quiet private area, a coffee shop, a room with an armchair . . . or a pub!

- It is about thinking of the other person's needs and trying your best to make them feel comfortable.

Timing

- *When* is it best to address the feedback? Now, or can it wait a bit? If wait a bit, why?

- Think about what you know about the person's schedule, so you can find an appropriate time—not rushed in between meetings or deadlines.

- Likewise, you want to be fresh, too, so think about when *you* can make space in your schedule so you don't arrive rushed.

Approach

- What's the opening approach you can take, one that best suits the person? Is it easygoing: "Hey, can I get a moment?" Or more direct: "Hey, can I give you some feedback/thoughts?" Some people may appreciate the boundaries of a more "formal" approach.

- Whatever approach you take needs to be based on compassionate thinking of who the person is and what suits them best to feel at ease.

- Remember voice tone. A frustrated voice tone creates fear.

2. It's All in the Understanding—Finding Shared Meaning

When we first met Max, one of our clients, he was off the charts on sarcastic and highly oppositional. In his organization he was perceived as the culture blocker who fought every idea every step of the way. His own peers in his leadership team were very wary of his sharp and cutting sarcasm. It was hard to read Max's behavior in any way other than a negative light. Until on a program with us, when we asked him, "What do you think is the purpose of using that behavior?" Max made an incredible statement that changed everything for him and for his peers. What he said was this: "I'm cynical because I *care*. I *don't want anything bad* to happen to our business." His intention was to critically evaluate everything in case something might be detrimental to the business. When he said that, people saw him in a very different light. He was expressing his true intention of his heart—one of intense loyalty and a drive to protect—but it was coming out as BTL because he had tied it to fear.

If we can find the purpose, the intention of a BTL behavior, we can find some degree of *shared meaning*. In this case, the shared meaning was that all the team were passionately committed to their business, and they all deeply cared about it. Once the rest of the team understood Max's *intention*, and then understood his BTL behavior, things changed dramatically. Max committed himself to lowering his sarcastic and controlling tone and asking questions to help with evaluation. His go-to message became "Help me get to yes." And he has become a very different man these days—an effective CEO in his own right.

Perception is everything!

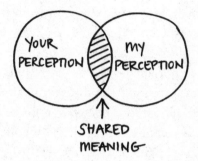

One of the most powerful words in the English language when we are wanting to develop people and use feedback is the word "perception," because it removes judgment. You come together saying "Let's just share our perceptions on this, what's your intention here, please help me understand where you're coming from, let's get on the same page together, can you please give me your perspective and I will share mine and hopefully get some shared meaning."

It's such a great way of giving feedback: we can be specific by giving examples and start by using the words "My perception is . . ." to make it safe for others to receive feedback and to share their perceptions. What happens then? We can peel back the layer of perceptions and reach the shared meaning that lies beneath.

Finding one positive truth about an initially negative situation or person is incredibly powerful because you're choosing to give up your tightly held sense of judgment and trade up to compassion. You get the opportunity to let go of your own standpoint! What we're doing here is giving up judgment in order to have empathy for another. The difficulty is that feeling "right" is so seductive—sometimes it's hard to let go of it!

This is the moment to stop thinking about the *behavior* they're exhibiting, and instead think about what their *intention* might be. In the workplace this might mean changing those previously consistently prickly interactions if you're both passionate about the common project you're working on—*that's* the shared meaning! Use that fact in common about you both to work as your foundation to get to an effective outcome.

At home this might mean stepping away from arguing with your partner or roommate about how much effort each has made in keeping the house clean (or not!). If you both *care* about the house (whether it's tidy or not) or about the relationship, one of those can be your shared meaning, not the disagreement about who does/doesn't do what. It's an important step to going ATL.

You can shift the circumstances through compassionate-based thinking and behavior that's not soft and fluffy—it's quite practical. And it works.

3. Separating the Person from the Behavior

This principle, old as it is, from the book *The One Minute Manager*, is worth mentioning again. To operate effectively in compassionate, the challenge is to practice separating the *person* from the *behavior* and then finding a *shared meaning* between us. "I value you as part of the team and what you do, and . . .; I love you son, that was an amazing game, and . . .; I love all you are doing, darling, and this one thing I would like to talk about . . ."

Nate, a senior technician in a medical lab, says the best piece of advice he ever got for addressing workplace conflict actually came from his high school soccer coach. It's just six words: "Play the ball, not the person." In soccer, if you attack an opposing player by shoving or kicking them, you'll get a penalty—and possibly injure someone. Go after the ball, because that's your real target. What that means for Nate is if one of his colleagues is missing deadlines, or taking credit for another teammate's successes, or any other behavior that needs to be challenged, he coaches by talking about what they're doing (the behavior) rather than who they are as a person. That might look like "We need to deliver this on time" rather than "You're always late with deadlines."

No one wants to offend anyone. We all know that when people are passionate about anything they do, feedback can be taken personally. So as much as we can reinforce that we respect the person, and we want to give them coaching on something they are doing, it can be difficult for people to not take things personally. When coaching someone we might be gracious by giving them the time to go to "night school": they spend a night or two (sometimes a week or two!) contemplating the feedback, and we then follow up with compassionate coaching.

Being effective in developing means being comfortable challenging others, and doing so by focusing on the issue—not the person. It means being known as someone who sees what people are capable of achieving and helps them work toward their personal best. Being comfortable and skilled at dealing with difference and conflict in a constructive manner, we can offer authentic encouragement and corrective feedback as appropriate. It's about being thoughtful and wise, waiting for "moments of truth" to happen before providing effective feedback that will be received as authentic and supportive. It allows us to communicate to others that each step of learning is to be celebrated, and mistakes are merely opportunities for improvement.

4. The Five Cs for Coaching

Because we want to develop people in both Heart + Smart, we have designed the five Cs for coaching matrix to help compartmentalize areas of development.* The document is used on a monthly basis when someone is new, then quarterly, biannually, and when relationships are *really* strong you can throw out the process! The team member fills it out and then it's discussed, with love and compassion.

- **Character: growing character.** What is happening for them, as seen in their Indicator and behavior? Are they actively working on building their ATL character, on growing the gold within their heart? How can I help them build their confidence?

- **Clarity: role clarity.** Is the person clear on their role, responsibilities, and expectations? Do they have the relevant information they require to effectively perform in their role? How can I help define further clarity?

- **Competency: building skills.** Does the person have the skills to effectively perform in their role? What training do they need? How can I support the person in getting the necessary know-how?

- **Connection: team player.** Is the person connecting with others, being a team member, honoring all people, and performance managing? Are they connecting with the clients/customers and going the extra mile? How can I help them see the benefits of connecting with others?

- **Commitment: passion.** What is the level of engagement of their heart in what we're doing, the business, the brand? Is it: 1) heart's not engaged, 2) growing heart engagement, 3) heart is totally in it, 4) losing heart, or 5) lost heart? How can I/we help them become more engaged and passionate about what we do?

5. The COACH Process

There are times in life as a leader, parent, partner, or friend you need to sit with someone and have a real heart-to-heart session.

English football (soccer) is a big deal in the UK. At age seven Daniel had started playing for a fairly competitive soccer team in his local region. Even though he hadn't trained, it was evident that he was a natural. For the first twenty-one games, his team, who had never played together before, won every game they played— Daniel scoring lots of goals—and his team got used to winning. They made it all the way to the finals.

In that final game, they faced a team that was also very good, had played together for a long time, and had won the cup for the last two years running. Daniel's team lost against them 7 to 1. It was the first game the boys had ever lost, and the first time Daniel himself had experienced *not* winning. He wasn't used to those feelings, and he couldn't cope with them. He threw down his water bottle, shouted at his teammates, walked over to his mom, and threw his jacket at her. What upset his mom most was Daniel not shaking hands with his own teammates, the other team members, or the other team's coach, who came to him to shake hands.

As a mom, Heidi says, "I was shocked and disappointed for him, but it made me realize this was a massive opportunity not just for Daniel to develop in his own character, but for us as his parents to learn how to help him in his character development." How easily competitive thinking and behavior can sneak up on you, even at seven years of age. Heidi's wisdom is that she knew to wait for an MOT.

One day Heidi was in the garden with Daniel kicking a soccer ball about. "We just started talking about what happens for him with the losing and how that makes him feel," Heidi recalls. "And it flashed across my own mind that I had a deep template locked away—about when I was in the first year at middle school; I used to captain the netball team. Our team was very good and beat all of the other schools in competition. We got to the very last game in the school year and we were expected to win the cup and league— all of it was riding on this one game. Very quickly in the game we went 1 to 0 down, then 2 to 0 down. Although I don't remember the final score, I do remember getting progressively frustrated, angry, and losing my head. I remember coming off the court after we

COACH	
CONNECT	Focus on the person and the relationship you have, connecting with what you want for them, with love and respect. Finding an MOT and ETA for this session will help create a safe environment. · Can we have a chat/can I please give you some feedback? I've been noticing a few things where I think I can help you—can we talk about it? · Let's go through your IDP (Individual Development Plan). · Let's go through your Heartstyles PDG.
OPEN UP	Ask questions to seek to understand the other's perspective, and gather the facts on a situation. Remember to *see past the behavior* to the heart of the matter. · What would you like to get out of today? · What is happening for you right now, what is your perception of . . . please help me understand/I'm curious about understanding more about the context surrounding [this situation]? · What situations or triggers cause you to . . . ? (explore ATL and BTL) · If coaching with the HSi Personal Development Guide: What surprised you, what do you agree with, what do you disagree with . . . ?
ADD VALUE	Give your perception, being truthful, authentic, factual, and compassionate. · I may be completely off the mark/this is only my perception, this is how I see it . . . When [xyz] happens I've noticed . . . It makes me feel . . . It makes others feel . . . The team perceives it as . . . It makes the team feel . . . The impact on the customer is . . . · It seems to me . . . Could this be true for you; is this *the* truth or *your* truth? · Are you exhausted, tired, frustrated, drained/what is this costing you?
CHECK UNDERSTANDING	Find the shared meaning or where you are in unity. · What have you understood and heard . . . ? · What I have understood and heard is . . . · Where do you see we have shared meaning?

HEAD FORWARD	Together, decide on how to move forward with specific actions and support.
	· Based on what we have discussed, let's work our solutions together/what ideas do you have/what could you stop, start, and continue doing/what would you do if you were in my situation/how could you achieve the same result in an ATL way?
	· How can I help you move forward/what support do you need from others to succeed? Perhaps you could consider . . .
	· If necessary, you as the coach, make suggestions/ solutions for their development, being clear.
	· How will you/we measure your success/what would you like me to hold you accountable for and follow up on with you?

had lost and shouting at my mom; I blamed my team members and didn't show myself in my best light. And I vowed I would never play again."

Thirty-six-year-old Heidi says, "I had forgotten all about that experience until I was kicking that ball around with my seven-year-old son in the garden and we were talking about *his* behavior on the soccer pitch. I said to Daniel, you have a choice: you can either learn how to cope with these emotions and use them to help yourself play *better*, or you can let these feelings of frustration and anger affect your game." So we came up with a little phrase that would help Daniel remember S+T=B when he gets frustrated or angry. *Lose your brain, lose your game. Keep your brain, win the game.*

Just recently he played against the same team that they lost the cup final to. They drew 1 to 1, and the difference in Daniel was obvious. He came off saying, "Wow, Mom! Look how we've progressed!" He shook everybody's hand. It was a pretty special moment for Heidi, and a special moment for Daniel.

Heidi coached her son by being direct with him but also having compassionate understanding of where he was at and what was driving those BTL behaviors. When she had a revelation about her own experience and behavior, she chose an MOT to talk to her son about her own experience and its outcomes, and ask him how he wanted to behave. She also helped him with phrases that made sense to him and that he could recall when he felt the old feelings come up. To his credit, even as a young boy, Daniel made the ATL choice to improve and focus on personal best, not a competitive winning mind-set.

In an important character coaching session, trust the COACH process to help you. It's done relationally with humility and love. On the previous pages is a concept guideline giving you some structure to work with.

Using COACH is a back-and-forth conversational flow—connecting with heart—not an intellectual interrogation! If you're using COACH with your kids or your spouse, how do you connect appropriately? Just to be clear—we don't mean literally get the piece of paper out! We mean using the *principles*! What's the sort of language and phraseology you would use at work versus at home? That "Hey, can we have a chat?" approach may work with a colleague, but saying, "Hey, can we have a chat?" might not work with your thirteen-year-old! ETA can help here, thinking about environment, timing, and approach.

Coaching Somebody Out of a Role

Compassionate is not just the compassion of holding on to someone; it also might be *releasing* them. If people are in the wrong function, sometimes they can become *dys*-functional. Sometimes people come along to do a job and they're not equipped to do the job; sometimes the job changes, the goal posts move. Sometimes people get promoted a level above their competency, which is known as the Peter Principle. People want the promotion, we give them the opportunity, and it can become too much for them to handle. So how can I develop in being compassionate when I need to performance manage someone out of the business? How can I terminate someone effectively and release them into their next season of life?

We believe the art of doing it well is basically rooted in respect and love. People sense when you have their best interests at heart, even as you are moving them out of the business.

We've also heard plenty of examples of people being fired by text message, via an e-mail, or face-to-face in a way that hasn't been handled well at all. What stops us from doing it effectively? Fear. Fear of being rejected. Fear of rejecting someone. Fear of hurting someone. Fear of what it might cost the business. So, we'll hang on and hang on. Sometimes that ends up costing us more than we anticipated—in emotional energy, financial costs, and culture costs.

Catherine was the newly appointed CEO of an online home furnishing company. In the role for six months, with a background outside of this industry, she was on a steep learning curve. Catherine was finding herself continually frustrated with a key member of her leadership team, Albert, who had been with the company many years. She felt they were just not gelling, and she found his work methods to be less than effective. Eventually Catherine took the step of meeting with her CPO with the intention of working through a process for either developing Albert or exiting him graciously out of the business. The CPO wisely wanted to fully investigate the situation and met with both Catherine and Albert to get their perspectives. She discovered that Catherine avoided interactions with Albert because they didn't gel, and when something went wrong on Albert's team Catherine would blow up. Albert had deep experience working in online retail, so his expertise triggered a fear response in Catherine, making her feel insecure about her technical knowledge. Conversely, Albert felt threatened by Catherine, because with her extensive experience in previous organizations, her qualifications, and her go-get-it leadership style, he felt inferior.

The CPO coached Catherine in taking the first steps to constructively communicate with Albert about their relationship and his functional capability. This simple plan consisted of first finding shared meaning between them, and then being able to share their perspectives together to reach a better level of communication from which to build on the relationship and functional issues.

In the first meeting Catherine had with Albert she ended up in an eye-opening conversation with him. They both agreed that their shared meaning was a passion for the business, and they conceded they were each coming to it from different perspectives: Albert from years of experience and Catherine from a new set of eyes and previous other industry ideas. As this shared meaning was able to relax the fraught atmosphere between them, Albert revealed that he was terrified he was going to be fired; he was having marital issues and he avoided Catherine because he was paralyzed by this fear of losing his job. He admitted his home issues were distracting him from being his best. Catherine was able to tell him that it was not her intention to keep him on edge; she just wanted to understand what was happening in the business and be able to help him make good decisions. This encounter softened Catherine's heart to Albert, allowing her to be able to deal with him with compassion while working on the functional issues that continued with him. Catherine told her CPO, "I don't want *anyone* to be afraid of me or afraid of losing their job every day—his behavior makes so much sense now." Albert also came to understand that his family issues were not the whole issue for him—he was indeed slow in adopting the new technical systems being introduced, and he was taking that stress home and "dumping" it on his wife.

To move forward, first Catherine and Albert agreed on a weekly no-miss face-to-face meeting to review the week just gone and the plans for the week ahead. They would discuss what was going well, what challenges to look out for, and how they would connect to resolve during the week. It was set up so that Albert oversaw the agenda—Catherine then didn't need to feel she had to know what to cover and Albert could have a leadership role in the interaction. Second, when something went wrong, before reacting, Catherine would reach out to Albert and explore, listen, and ensure she was clear on the situation. Then she would ask Albert, "What is happening for you and/or what do you think is at the root of this situation?"

In the end, Albert did exit the business. He and Catherine developed respect for each other, and they both gained an awareness of some of the unstated issues that had been wreaking havoc between them. Importantly, they set up a structure that helped them overcome their desire to avoid each other while they worked on building their working relationship. Albert was able to feel that he finished well, having developed an effective hand-over plan, and supported Catherine in the recruitment for a new incumbent for his role. Catherine learned how important it is to find shared meaning and get to the heart of someone's behavior, rather than just reacting to them. Albert's personal life was better for it. He found a role that fitted his capability, leading to less stress for him, and this in turn had a positive effect on his marriage.

The Bookends of ATL: Authentic and Compassionate

Compassion gets a bad rap at times. Some people think of compassion as being a wishy-washy, "I love everyone and I get trodden on" sort of quality. Take another look at the ATL styles in the Heartstyles model. On the far left is authentic, and on the far right is compassionate. It's no accident that these two styles are our ATL "bookends," holding all the other ATL behaviors securely in place.

With authentic and compassionate working in a pair, we can set appropriate boundaries and have the humility and courage to speak important truths without judgment.

On the surface, it might look like authentic and compassionate don't have a whole lot of common ground; however, they are closely linked when it comes to not judging others. Authentic people who are comfortable in their own skin do not judge others to get self-worth. Compassionate people look past other people's behavior to understand what shaped their lives. I can be authentic in telling someone a difficult truth, yet compassionate because I have objectively discerned the information and I'm not going to judge the person.

Mara learned this in her career, in which she spent more than eighteen years as a forensic neuropsychologist/expert witness for murder trials in Australia. Her job was to perform neuropsychological assessments on prisoners awaiting trial as part of the trial process. She would come to each session with a huge kit of tests, because in assessing the brain you need to work through a whole process of "If this, then that. If not that, then this." It's another kind of puzzle: you've got many different tests and you've got to see how people react, because that will determine the next thing you need to test for.

Along with the neuropsychological assessments, you're also observing the person clinically. You can't have in your mind, "This person is charged with killing three people," because that will just interfere with your clarity and objectivity, and you could be in danger of drawing incorrect conclusions or making assumptions, Mara describes.

Mara says, "Part of compassionate is assessing: you're trying to work out what the whole picture is of this person in terms of their background, the mitigating factors, and so forth. That doesn't mean you're overly sympathetic to this person as such. It just means you might understand perhaps *why* they've gotten to this point— because of their history, or because of some brain impairment or psychological issue. But it doesn't necessarily then mean you're going to feel sorry for them. It doesn't work like that."

Although that's an extreme example, it's a similar principle to what we're doing in everyday life, which is why we say you can forgive someone without necessarily condoning the behavior. When you're operating in compassionate, there's that authenticity piece giving you an objectivity about yourself, and yourself in relation to this person, and this person in relation to a bigger context of their life. That way, you can understand how best to operate with this person. It doesn't mean you overidentify with them. It might mean you might soften your view because of knowing the context, but even that's a conscious choice. As Mara says, "There aren't many people in prison who had a happy childhood."

Having a sense of ourselves as who we really are and being confident in that, while at the very same time honoring others as important, too, is the kind of balance that brings fulfillment. So, the sense of boundaries also has a lot to do with authenticity. Sometimes people who are overly compassionate with a lack of authenticity have few or no boundaries and can get trampled on.

Motivation—Motive-Action

The word "motivation" is made up of two words—"motive" and "action." Having now read the past four chapters, you will have a deeper sense of the eight ATL behaviors and what they look like in your own behavior and that of the people around you. We now encourage you to examine your heart's motives and take action toward living and leading ATL. We invite you to take a few minutes to sum up in your journal what you can do to increase these styles in your own life, and what triggers, templates, and truths might be blocking you in particular areas. Here's a template for you to start with:

	Three actions I can take to increase my effectiveness in:	Are any triggers, templates, or truths blocking me in this area?
CH 7 – KNOW WHO I AM	1. 2. 3.	Trigger/s: Template/s or: Truth/s:
CH 8 – KNOW WHERE I'M GOING	1. 2. 3.	Trigger/s: Template/s or: Truth/s:
CH 9 – CONNECTING WITH OTHERS	1. 2. 3.	Trigger/s: Template/s or: Truth/s:
CH 10 – GROWING WITH OTHERS	1. 2. 3.	Trigger/s: Template/s or: Truth/s:

If your spirit can stay in compassionate and discernment, not in judgment, and be authentic, you've got a much better chance of coming across with growth-driven love. Once again, it's no accident that developing is sitting alongside compassionate, because this is where the developing part comes in to partner with compassionate. With authentic and compassionate working in a pair, we are able to set appropriate boundaries and have the humility and courage to speak important truths without judgment. As Mara puts it, "The more authentic a person I am, the more I have built my character and the more I can be compassionate to the world, as strange as it may be at times!"

A Final Thought: Tears

Tears in life are a beautiful expression of the heart. We have happy and sad tears, and as we go through life's journey, we can experience both. So often, though, we are afraid of tears. Men have been taught "big boys don't cry," and women have been accused of being overly emotional. Yet one of the beautiful languages that we all have in common is tears.

Tears speak from the deep well of the heart. They have a voice that can release the expression of joy and unlock what's been holding people back.

Have you noticed when someone achieves something they've worked hard for, wins a gold medal, or is paid a compliment and recognized for their achievements, they well up with emotion and maybe even cry? These can be achievement tears, recognition tears, happy tears, relief tears, release tears. Yes, there are also sad tears, hurt tears, tears of rejection, tears of disappointment, not-good-enough tears, exhaustion tears. The question is: What is behind the tears?

When we're making a sincere connection with someone—be it colleagues, friends, family, children—and tears appear, a compassionate, wise response can be to ask, "What *sort* of tears are those tears—happy or sad tears?" This gives the person a simple starting place to respond from, to share their heart and what is truly happening for them, rather than us making assumptions. Stephen recalls an aha moment more than twenty-five years ago, when he was coaching someone and they burst into tears, saying, "Thanks so much for that feedback, I feel so relieved!" He realized then that there were different *kinds* of tears.

Let's be courageously caring for one another as we courageously coach others, even through the mix of emotions that life can carry at times. It can be a wonderful MOT when people share their tears and their heart. It is truly caring and can help people clarify what's happening for them in their growth journey. The heart speaks through tears, and if you have a safe place for people to share the deep things of the heart, even with tears, you are truly connecting with others.

❤

CHAPTER 11

Building Above the Line Work Culture

If you met Hugh and spent an hour or so with him, he might not be the guy you would think of first to rally the troops, or lead the charge, or motivate and inspire an entire company of more than eight thousand people spread across a country. He's an introvert and not a fan of public speaking. So when he was hired to lead the turnaround of a restaurant brand that had been declining for a decade, it might have been a surprise for some. But over the last few years, he and his team have transformed a business that otherwise would have been gone by now. The look and feel of the brand have been refreshed through the tangible elements of product, menu, and decor changes, and the company is reclaiming market share. But that isn't the whole story of how they got to where they are.

So, what happened?

What Hugh saw immediately on taking the CEO role was that he would first have to turn around a toxic culture. A culture that had become so *below the line*, so stuck, so "dead," it needed a radical transformation, and someone had to take the risk to change it. Hugh did.

He could see that the fear of more financial losses, including the potential for lost jobs, and the pride of not wanting to be seen as partly responsible for poor performance were leading to dysfunctional behavior. Managers trying to control every aspect of employee behavior, office politics of blame and avoidance, ineffective decision making that caused everything to take twice as long as it should, and leaders frantically jumping from one crisis to the next. Many people were disengaged, keeping their heads down, showing up for the paycheck and not much else, believing the company might collapse at any moment. The high-performing, talented people were being headhunted out of the business.

How would a polite, quietly spoken leader overcome these negative aspects of the culture? Hugh is a wonderful combination—both extremely well-read and educated, and a man whose personal values and behaviors were very much driven by humility and love. We saw this in him as soon as we met him, not long before he took on this new, difficult role. Those principles have helped him grow and achieve beyond his own fears and discomfort with being in the limelight.

Love-driven relating and compassionate helped him to understand the staff needed to feel cared for, supported, and encouraged in their efforts. Employees needed to believe in the company again, but to engage them in that way, the leaders needed to show they believed in the employees.

Through humility, Hugh had the inner strength and courage to stay committed to what he knew was the right thing to do, regardless of the immediate financial needs of the company. He knew who he was and where he wanted to lead the company, and he wanted to do that *alongside* others, not *in spite of them*. He carried a spirit of openness and integrity, a sense of confidence, and especially a desire for personal growth. Hugh radiated this authenticity and integrity and desire to learn and improve and used it to create a new positive tone within the company.

What, specifically, did Hugh do? Once a week he would meet with the leaders and pair transparent updates on current finances with inspiring stories from around the company of great customer service or a restaurant that had turned its numbers around. He was an authentic cheerleader, and his calm command and inspiring messages gave people hope. Every Wednesday night for a year, he went to one of the three hundred restaurants and worked alongside the staff as a server. He wanted to understand what their work was like, what was most difficult for them, what interactions with customers were like, and what systems and processes were blockers or broken. He could meet servers, model the behaviors they were trying to grow throughout the company, and then talk about his experiences with other leaders. (On his first night, he was so excited to receive tips from customers, feeling what it was like for the team when they performed at their best.)

Hugh knew that to get the results, he and his people needed to revise all the business practices—from product, service, decor, menu, systems, and the employee experience process. He knew he needed total engagement from his people, and the way to create that was through great leadership and great culture, so that is where he started.

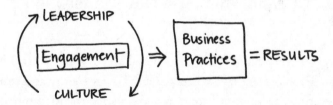

He worked to free people from the fear of failure—and thus speed up decision making and action. Jokingly he would say, "Let's try something new. If it doesn't work, we can go back to what wasn't working anyway!" and then in a serious way, "If it doesn't work, we will come up with another idea." When things went wrong, he wasn't punishing in his demeanor or language. He was compassionate, but focused on solving the challenge and achieving something better—quickly. "Let's talk through what happened," and he then listened closely for what they could learn from the experience. And while he had to overcome his own fear of speaking to crowds, his authenticity and integrity made him a more powerful and inspiring speaker. His authentic belief in the company and in the employees meant he was able to connect deeply with others and draw out their true talents, ideas, and visions.

Our friend author Tommy Spaulding would say it took a "Heart-Led Leader" to see what the company needed most and create it. Hugh made people feel safe and hopeful and cared for—because that's what it's like to be around people who operate above the line more often. As a result, people trusted him, believed in him, and wanted to work alongside him in resurrecting the brand. The incredible rise in heart engagement and passion for doing good work throughout the company was proof enough of the transformation he had led, and the incredible financial success was a wonderful outcome.

From Hierarchy to Partnership: Leadership for a New Generation

Have you considered what energy and spirit you are creating as a leader? Even better, have you considered what energy and spirit you are creating as a *human being*? When asked at the Melbourne Press Club (Australia) in 2000 which leaders of the twentieth century he admired, Nelson Mandela was adamant. "It's not the question of a leader. It's the question of a human being who does something to make an ordinary individual feel, 'I am a human being, and I have a future, and I can go to bed feeling strong and full of hope.'" Every one of us has the capacity to be a leader in this way, investing our hearts and minds in helping our brothers, sisters, children, neighbors, tennis partner, corner-store owner, bus driver feel strong and full of hope.

For our grandparents, a good workplace culture might have been a place where a team of people silently got on with their jobs, completed tasks, obeyed rules, and clocked off at their assigned time. Old-style leadership was based on *a hierarchy of pride and fear.* In the world we live in today, though, value is placed on *a partnership of humility and love.*

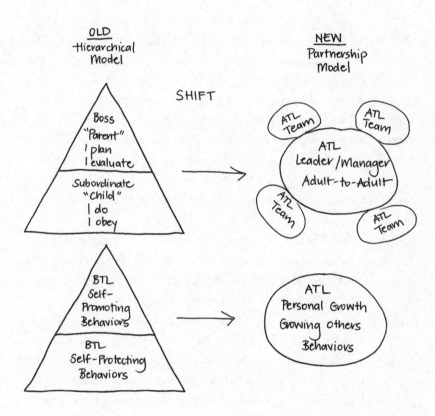

The Industrial Revolution was built on a hierarchical model with premises such as these: *I'm the boss, you're the subordinate. I plan, you do; I evaluate, you obey.* In this model, employees could come to work and not have to think beyond accomplishing a set of tasks. This was the *production line* mind-set, driven by what we call parent-child culture. Now our society has matured—where baby boomers wanted control, millennials want community. A recent report from Gallup summed it up by saying that millennials want what previous generations wanted: a life well-lived, good jobs with thirty-plus hours of work a week, regular paychecks from employers, *but* they also want to be engaged (emotionally, intellectually, and behaviorally), they want high levels of well-being, a purposeful life, active community, and social ties.

Now we've arrived at a generation that is highly educated, so the gap between *us* and *them* has vanished. In fact, you've probably found yourself in a situation in which people younger than you know far more than you, as a leader, about a certain subject. They're savvy about concepts and technology. (Need a lesson in using that new software? Grab the nearest teenager and ask them!) The gap between those hierarchical levels has broken apart, and yet some are still trying to run organizations that same old way. When we shift over to a partnership model with its circular relationship in which we plan, do, and evaluate *together*, that's when we can channel all that energy of young people and their ideas, awareness, and knowledge.

The truth is that an ATL culture eliminates the need to worry about generational differences. The way people of individual generations interpret or relate to ATL behaviors might differ slightly. Above and beyond those differences, though, we don't see any real difference between generations. Every generation has wanted authentic, transforming, reliable, and achieving people who are relating, encouraging, developing, and compassionate. Now, organizations no longer need to think about engaging different *generational* needs. When you're ATL, you're engaging basic *human* needs.

Effective leaders today breathe life into the environment by making it a safe place, so that people can express opinions and ideas and disagree in robust debate without fearing they will be punished. In a healthy culture, the ATL leader will have the humility to admit any wrong decision quickly, evaluate together with their teams, and be nimble in turning the ship around. After all, organizations don't innovate, people do. Organizations don't turn a business around, people do. *Organizations don't lead culture, leaders do.*

You Cannot NOT Have a Culture

Here's the thing: *you cannot NOT have a culture.* Put three people in a room, in a team, a business, a family, a sporting team, and boom: you've got a group dynamic, and so you've got a culture. At its most basic, culture is where something grows. In a healthy culture, something grows well. In an unhealthy culture, something cannot grow to its potential; instead, it becomes fragile, unsustainable, and toxic—weeds come in and eventually it dies. As complex—and at times baffling—as the human body itself, culture is created, developed, and impacted by countless factors from both the inside and the outside.

What Creates Culture?

Many aspects of culture are shown in this illustration, but our focus is *behavior.* In the past few years, you've probably seen organizations spending huge amounts of time and energy trying to build an appealing culture through office refurbishments, bonuses, parties, team-building programs, pool tables, baristas. All these things can be great, but if behaviors throughout the organization are not primarily ATL, the enticements lose their shine pretty quickly. Our generations are savvier and more cynical than ever, seeing straight through attempts to bribe them into happiness, while at the same time yearning for more meaning and true connection in the workplace.

Leaders and managers everywhere these days have culture development as a KPI (key performance indicator). It feels like foreign territory for many, though. "I went to college to become an engineer/lawyer/accountant, not a psychologist!" This responsibility to drive culture can feel like just one more thing to manage: years ago it wasn't really on our agenda. Or if it was, it was just a matter of making sure the fridge was stocked with beer for Friday happy hour. Some companies are constantly changing how they do things in the hope of shaking up their culture. Yet without a long-term vision and an intention to create a courageous, honor-driven culture, it can all just be floundering and chaos.

Hits and Misses: CEOs on Culture Development

The million-dollar question, then, is this: Why should an organization invest in culture and people development? It might sound like a good idea, but what kind of impact does it really have? And how do you set about developing your people in a way that will truly have the kind of positive effect you're seeking? (Yes, that was really three questions, not just one. But they all point in the same direction!)

Alongside Hugh, we asked five of our clients—a global CEO, national CEOs, and CPOs from successful and respected businesses—those very three questions about their own experiences of investing in culture and people development. Each of these leaders had led their businesses through a process of transformation using ATL principles. Each of them had seen concrete results in staff retention, increased profits, and customer satisfaction. We wanted to hear from them *why* they set out on this journey in the first place; *how* they set about it; *what* results they saw; and what lessons (or face-plants!) they encountered along the way. We have changed their names to keep confidentiality.

Q1. Why did you invest in culture and people development—and keep investing year after year?

All of our clients agreed that sound processes are one thing, but their execution is wholly dependent on people—their mind-set, their attitude, their teamwork, their support. All six reported they wanted a culture where people feel supported and empowered and had a voice, as motivated people are generally more productive and need less supervision or instruction. "Skill competencies ensure the 'what' gets done, but it's character that ensures the 'how' happens in a way that is motivational and inspirational to the broader business," explained Lucy.

Hugh added, "We've always been skeptical about worthy sets of values and mission statements that were really personal beliefs of our senior leaders. We wanted to build a culture where the values helped our people do life outside of work. We are in a minimum wage industry, so we have a duty to create a work environment where people feel valued, have ample opportunity for personal and professional growth, and are supported during difficult periods in life. We also wanted a strong culture to win the war for talent. It's easier to retain and attract the best."

Shane shared his story of more than twenty years as a CEO.

"For us it was prompted by a very specific moment in the business. I decided to measure culture after some very disruptive events in my family business. Around that time, I was overseeing some very high staff turnover, discovered some serious gossip and manipulation, and found opposition to ideas and some other symptoms of low team engagement. I genuinely wanted our brand to become a household name. So, we undertook a culture survey and were shocked by the poor results. Consequently, we made changes in the business. Some were significant and immediate, addressing obvious shortcomings, while others required deeper examination and consideration. The changes took some years as we continued to unpick our systems and practices. Since then we've continued to measure culture every year, and recently passed twenty consecutive years of culture surveys. Annually we measure our P&L and our culture. Deciding to continue each year to assess our culture is embedded in our ambition to be a famous household name."

Q2. What outcomes have you seen in your organization?

Shane's story is very inspiring for its concrete results. "The numbers were undeniable. In the years following that first culture survey, staff turnover dropped to below 20 percent (previously it had at times been over 50 percent). Implementing change became easier. Revenue and profit growth were strong. We acquired our biggest competitor in a multimillion dollar deal. That was an . . . interesting experience. Our new constructive culture collided with the very aggressive dog-eat-dog culture of our competitor—their fierce enemy was now their owner! Success in our competitor was driven by win at all costs, even against other branches, don't question the leaders or supervisors, even cover up for each other in taking time off—a very 'us and them' mind-set. This was now totally foreign to us. The merged cultures did not go very well. By the following year, there was a clear retreat in our culture results and we were paying a price. There was unhappiness and staff turnover spiked. But as time went on, trust grew, our organization's ways endured and were embedded across the new parts of the business. Two years later our culture survey results were back in a strong place. The takeover turned out to be a huge success and a real game-changer for us."

Lucy shared a similar experience: "We also had a difficult time in our business, and the most significant outcome for us was being able to engage and retain talent during that time. In addition to tough economic headwinds, we experienced significant transition in leadership. The culture we built allowed us to keep talent motivated and inspired. We were losing market share, then regained it—clearly through our people and culture. The multiple different types of learning programs we put in place provided structure, shared language, and measurability to the way we build culture. We got engagement that turned into increased sales. That created retention."

"Unity of purpose, working effectively together through thick and thin, and maintaining a high-level enthusiasm at the front line were all essential," Hugh commented. "At an individual level, countless colleagues have commented how the programs have not only changed the way they behave in the workplace but have also impacted their lives more broadly, including their relationships with family and friends. Both of these outcomes play to the purpose we share as a top leadership team: to create a commercially successful business that, at the same time, provides an environment for colleagues at all levels to feel respected and nurtured, to grow and develop."

Hayden added, "We've also experienced lower staff turnover, so we've been able to grow from within—most promotions have been from inside the company, saving on recruitment and training costs. We're also seeing employees who care, great customer satisfaction measured through our net promoter scores, a flatter organizational structure, growing top line revenue, consistent great bottom line results, and a strong balance sheet."

Anthony had an interesting point on the workplace dynamic they now see: "This sounds contradictory, I know, but our experience is that now that we have a strong culture, people are being more demanding of each other. People focus on the issue itself, without having to do the awkward dance around whether they will offend someone, or someone will take them the wrong way. The elephant in the room is discussed. No more gossip. If people know that the other person is not out to score points, or undermine them, or take inappropriate credit, then both parties can cut to the chase and focus on the problem at hand. People can be real, speak their truth—critical when they need to be—of the work done by others, because the other person appreciates there's no malice, just a genuine desire to get the best outcome. People build on top of other people's successes, rather than dismiss, or tear things down. Our meetings are more efficient and effective and robust. Decision making is discussed, mistakes admitted and fixed quickly, and we have become more innovating and agile."

Q3. How did you go about creating an ATL culture?

Our final question was probably the same one you have in your head right now: *How* did you do it?

Shane shared some great learning. "We started by turning the mirror on the entire leadership team. They illustrate what is tolerated and what is rewarded. Because we were dealing with a crisis in culture, I took a more radical approach. In the first instance my small leadership team was disbanded with each of them offered to return to their old role, and to reapply for the team or to leave. No one from the old team was returned to the leadership team. On reflection, these were good, well-intentioned people who just didn't have the capacity to lead teams. They were in the position because of technical skills and tenure but underdeveloped for culture and people skills. My fault, not theirs. It worked out they were much more comfortable being led than leading and got a hold of the new culture. I started again and rebuilt the team, hiring for culture.

"Concurrent with the leadership change, we released the survey results to the entire company. We engaged in discussions with everyone about the results, including my own candid disclosure of my role and shortcomings. We needed to change the perception that head office had agendas and secrets. We had to get honest, and it started with me.

"Beyond this, we did many other things to remove systems that encouraged the wrong outcomes; policies and practices that didn't support our direction. Established practices were challenged. If they didn't serve us, they were removed or changed.

"Our pay and reward systems were changed away from win/lose to win/win. No more bonuses that rewarded first place over second place. We set our KPI rewards to attainment levels of gold, silver, and bronze. And we wanted everyone to be rewarded at the gold level! Idea sharing between teams began to flourish.

"We also looked at the hard edges in our customer-facing practices. We removed a range of charges that had no real value, but put our people into uncomfortable discussions with customers. These business rules speak volumes to your people about the type of company you are.

"Beyond this low hanging fruit, we unpacked our business to find the source of what was holding us back. We mixed it up. Tried different approaches. We held road-show focus groups. We sought written responses and, in some years, we also held conferences to consider culture—and those conferences were designed and led by the employees."

Lucy summarized their approach: "A partnership with the general manager of the business is what got the momentum going on the leader development component of the bigger culture journey. Ours was an inside out approach—we did this because we believed effective leadership is only possible when we connect with people's hearts. It was positioned as an investment in personal growth with the intent to fuel results, because people would be bringing their best selves to work.

"Leaders with character also have the confidence to do what is right in those times of uncertainty and ambiguity—that's pretty much a constant in our Africa operations, which is my main area of responsibility. Building in plenty of time for our leaders to reflect on character and how it manifests in behavior—through leadership development programs—means they tend to show up consistently over time. It's this consistency that drives trust and respect with direct reports. The outcome is employees who are willing to follow even when times are tough.

"The biggest lesson was to ensure the language became a part of the DNA of the business. Doing this as a once-off program is not going to drive long-term success in your business. Today, character conversations are built into every aspect of our people capability cycle—individual development conversations, performance reviews, and talent planning."

Hugh reports his experience: "With my leadership team we adopted a saying 'Don't blame them, train them.' That's where the work started. Internal employees were chosen to be certified to facilitate personal development programs. The trust grew that the programs were for development not assessment, and momentum started happening. Enthusiasm, energy, positivity was going viral. Managers caught the message and took it to the front line.

"Once we had created momentum and learning around the principles, common language of above and below the line, and understanding of the sixteen behaviors, we took a radical approach by a) anchoring our culture in above the line behaviors, b) measuring our culture through completing an annual Indicator for the top five hundred in our business, including restaurant managers, and using this as a foundation for ongoing personal development and coaching, and c) using improvement in the aggregated Indicator data for the total business as a key performance metric, making culture measurable and more actively manageable."

In Anthony's case, the work also began with the leadership team: "It starts with building self-awareness using a 360-degree tool for senior leaders who are prepared to walk-the-talk, role-model vulnerability, and focus on trying to behave more constructively. Having senior leaders tell their own journey stories and allowing regular personal sharing makes it completely okay to be vulnerable. It demonstrates that everyone is in this together as people, and that senior leaders are on their journey, too.

"Second, you then allow more and more people to get exposure to tools that provide them with the skills to become more self-aware and appreciate the impact of their behavior on others. As more people become self-aware, they see those who practice constructive behaviors doing well, and succeeding; a common language also develops. Leaders demonstrate that the outcome is far more important than who is right, who is more senior, and actively acknowledge and reward people who are collaborators and can work in a team, as much as those who come up with a great idea in the first place.

"This fuels a desire to not only grow and become a more effective ATL leader, but also to be a positive contributor to the broader culture of the organization—because they see the multiplying benefit of constructive interactions at every level, across all aspects of the organization."

Five Levers for Organizational Change

Going by the experience of these CEOs, in nations across the globe, it seems that culture is the new frontier of a competitive advantage for talent and client retention. The next revolution in business is what we call the Heart Revolution (HR renamed!). To be part of this revolution, business leaders need a culture-focused way of thinking and a heart-based set of tools. The key is to have disruptive thinking within teams, who innovate and question the status quo *before* a crisis comes to be the catalyst for change.

Make a Case for Culture

In their own organization and that of their competitors, Shane and his team have seen that what *challenges* the business *changes* the business, and most organizations leave it too late to steer those changes in a positive direction. Yet Shane's story shows us how a toxic culture can be transformed with courageous and loving leadership, and how significant the business outcomes can be. AI is here already and growing quickly. With a changing online world and every industry being disrupted, there is a case for "what got us here won't get us there," to paraphrase business educator Marshall Goldsmith. This is why we say it is part of every leadership remit to develop a company culture strategy.

The first stage of that strategy is simply naming your culture. Are you about innovation or honor? Known for your capacity to rejuvenate and reinvent? Particularly progressive or nimble? By clarifying and explaining the key elements of your organization's culture, you will provide a beacon and goal for your people to work toward.

From there you can design a model to put it all together, combining:

- What the organization stands for—its vision, mission, purpose

- The expectations of your organization's culture, its values and behaviors

- A plan for how the organization will develop people through learning and training

Can you see your leadership team being deliberately developmental to create a commercially successful business that, at the same time, provides an environment for employees at all levels to feel respected and nurtured, to grow and develop? Considering that many of the frontline employees in organizations have never had access to this level of coaching and support, creating a case for culture development that incorporates *every* level of the organization can certainly be a way of leading in the marketplace for the gaining and retaining of talent.

Explain to the entire organization why you all need to develop culture, communicating what's in it for the front line. Being able to define a culture based on universal values and making it relevant and actionable at an intimately personal level for everyone in the business will mean true ownership for growth.

Measure to Manage

If you measure it, you can manage it—or to take that one step further, what gets measured in business gets managed. To measure "what it looks like," the outcomes of the business—from financials to engagement—are obviously important. To also measure "where it comes from," the behaviors of culture are equally important—measuring the *heart* of the culture. When people complete the Heartstyles survey (in chapter 4 we explain how to do that online for free), they are asked seventy-five questions to get the Indicator results, and then eight further questions regarding outcomes. All the results are shown in the participant's PDG (Personal Development Guide). The respondent's data can be aggregated to see if there is a correlation between ATL and BTL behavior and the eight outcome questions (based on the participant's perceptions). Following is an example of four of those outcomes:

1. Effectiveness at work

2. Personal development

3. Relationships

4. Time management effectiveness

Current level of effectiveness at work

Respondents who scored participants **very low** to **moderate** (n=21117 : 17%)

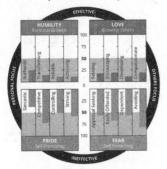

Current level of personal effectiveness

Respondents who scored participants **very low** to **moderate** (n=22858 : 19%)

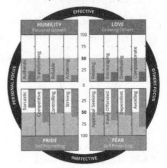

Respondents who scored participants **high** (n=54332 : 45%)

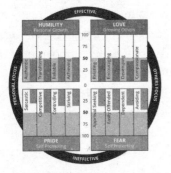

Respondents who scored participants **high** (n=61254 : 51%)

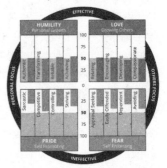

Respondents who scored participants **very high** (n=45474 : 38%)

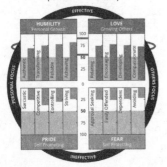

Respondents who scored participants **very high** (n=36811 : 30%)

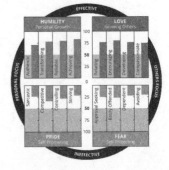

Current level of quality of relationships

Respondents who scored participants **very low** to **moderate**
(n=26525 : 22%)

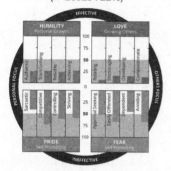

Respondents who scored participants **high**
(n=54845 : 45%)

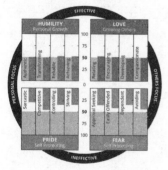

Respondents who scored participants **very high**
(n=39553 : 33%)

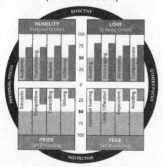

Current level of time management effectiveness

Respondents who scored participants **very low** to **moderate**
(n=34964 : 29%)

Respondents who scored participants **high**
(n=51061 : 42%)

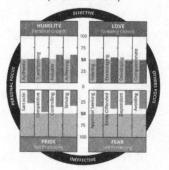

Respondents who scored participants **very high**
(n=34898 : 29%)

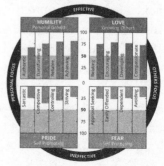

The above examples consistently show that when there are high levels of ATL behavior there are very high levels of outcomes. Conversely, high levels of BTL behavior are linked to low to moderate levels of outcomes.

ATL in the DNA—from CEO to Front Line

Have you ever been in the room when this conversation is taking place?

CFO: What if we train them and they leave?

CEO: What if we don't and they *stay*?

To get ATL behaviors into the DNA of an organization calls for an all-in-it-together approach, from the CEO to the front line, with *everyone* taking responsibility for developing the culture—not just the leadership. The gas in this particular tank is the attitude "If it's meant to be, it's up to me." The credibility that is born when the senior management team talk about their personal development journeys and are open and vulnerable about their own development opportunities creates a one team, one direction culture. As we have seen, this kind of humility is not weakness: Patrick Lencioni captured it well when he wrote, "At the heart of vulnerability lies the willingness of people to abandon their pride and fear, to sacrifice their egos for the collective good of the team."

It's common knowledge these days that there is a desire, even an expectation, from our current generations that organizations provide personal and skill development programs. Since we started our business back in 1987, our programs have essentially been personal development programs, even though we have called them by different names. We have seen how being intentional about providing personal development by an organization—from the C-suite to the front line—unlocks loyalty, honor, respect, and engagement in people for the brand and the culture.

We have all seen how *behavior can be a culture builder or a culture buster.* For the individual to create engagement we refer to our simple equation:

Heart (personal development) + Smart (functional skill development) = Engagement

Engaged people then drive business practices and thus results.

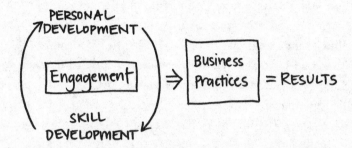

It's as predictable as gravity: where there is low trust, people will self-protect and self-promote, and this is how we end up with a BTL culture. The politics of corporate life manifest and drive culture, and training programs alone can't and don't overcome those influences. When there is a shift to ATL behaviors, trust grows, and when trust grows, there is a shift in ATL behaviors. The vital element is *consistency*, the kind of *sustainable* trust that creates ownership and commitment (as seen in Hugh's story).

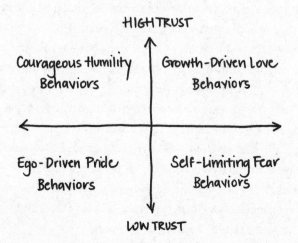

The investment in giving people a compass for life is also investing in furthering their ability to navigate life. Mental health, well-being, and resilience have become priorities in the current world of work. We expect staff to be engaged in their job when they're not engaged in life, but the ability to effectively deal with ups and downs and bounce back from challenges is important for productivity, and can be applied to employees' abilities to manage anything from a tough workload to frustrating colleagues. It is said that those with greater resilience are better able to manage stress, which is a risk factor for conditions such as anxiety and depression. Resilient people are also flexible, able to adapt to new and different situations, learn from experience, are optimistic, and ask for help when they need it. These are all ATL behaviors that make people more effective human beings in all areas of life.

Systems and Practices

In Hugh's business, they changed many of their systems and processes to encourage a more ATL outlook. "League tables" in their restaurants were abolished, operational practices became less control-driven and "approval seeking," stylized PowerPoint presentations to the leadership team were replaced by more factual one-page documents. In their new culture, recognition programs also specifically call out and reward ATL behaviors.

Reward systems, IT systems, recruitment systems, and communication systems can all have a *negative* impact on culture. Some impact is obvious, but some is very subtle. Shane's point about their traditional reward systems carrying on a culture of ego-driven competitive is a good example. They went to the granular level of examining all their reward systems and revamped them to ensure that they were motivated by an ATL achieving mind-set, not a BTL competitive mind-set. This kind of granular examination of the way an organization's systems operate is warranted and necessary for real and sustainable culture development.

New systems can bring their own set of new problems in the beginning, as Kristina found. "As a developing organization, we are still prone to seeing challenge as rejection and going through the cycle of approval seeking, striving, easily offended, before it rises up and galvanizes into achievement. We can also be a highly avoidant culture seeking parental direction. For example, the removal of some core well-imbedded past systems was initially taken as an excuse to take our foot off the pedal, and performance went backward in the short term." Seeing the slide, Kristina's team took action. "We have created new, more effective systems to mitigate what we know can happen when pressures build and people can start to go BTL."

Storytelling

Storytelling keeps culture alive. It is how histories are passed down, how customs are shared, and how traditions become endemic to a group. Success stories, recognition stories, stories of the past fun, struggles, and breakthroughs. Sales and client stories. Personal development stories and family stories. If you think back to the neuroscience information now available to us, you'll know how strongly our brains respond to storytelling: movies are the classic example (remember in chapter 2, we saw how deeply we react to what we know is just a story on the screen?). How even more powerful it is when we share and listen to real stories from each other, through which we can all become a part of the joint history of our organization.

A good practice is to set aside time to share significant events, both personally and professionally. One of our team calls it RQ— relationship quotient. This can be done over a team lunch, with each person getting time to share stories. In these storytelling situations, you have permission to be real! We need permission to be real, and yet our prevailing work cultures of the past did not necessarily give us permission to be humble, vulnerable, and love others.

When the CEO and all senior leaders give themselves and everyone around them *permission to be real* and compassionate while driving results with purpose, the culture can shift; individuals can become their authentic selves. On every level, that has true power.

As we were finishing writing this book, a CEO of a multibillion dollar organization we have worked with for more than twenty-five years was also finishing his career and retiring. We had a celebration catch-up with him, and he shared this meaningful story. "We had a great sendoff party at head office. In all of the speeches no one thanked me for all the money I made the organization, no one thanked me for all the great strategic decisions I was part of. All they did was thank me for the great culture I was part of building and the relationships I made, and how I cared for all the people in our organization—from the senior leadership team to the front-line fifteen-year-old worker!"

What a great story. It reminded us of that wonderful Maya Angelou quote: "People will forget what you said, people will forget what you did, but people will never forget how you made them feel."

The great leaders we have watched build great cultures and organizations have been aware that the language of business is money: no money, no business. They understand that strategy, structure, systems, and results are extremely important, but they also are aware that their role is *beyond* the task, beyond the money. They are deeply aware of what underpins the sustainable results. They understand it's culture.

These leaders know their leadership shadow is communicating energy and has the ability to change the atmosphere of the workplace. They focus on ensuring their *intentions* come across with an ATL *impression* to others, making their *impact* a positive one on the world around them. They know if they can radiate a safe place of *we're all in it together*, then people want to belong, then they can believe and thus behave in a way that adds value to the culture. These leaders ask themselves "why and so what"—why are we doing this and so what if we stay the same or change? They practice "helicopter leadership," where they continually rise above the day-to-day and hover, looking over the business and seeing where they can land and assist. In their "helicopter time," they can carry big loads of problems that need to be addressed, but they also know if the load is too large and too heavy there will be a crash. They lighten the load through effective delegation.

These leaders understand the "beyonds." In business we are often tempted to trade purpose for profit, but courageous leaders go *beyond* to create heart engagement . . . purpose *beyond* profit, meaning *beyond* money, commitment *beyond* convenience, destiny *beyond* daily, to unlock in their people passion *beyond* pay, service *beyond* self, identity *beyond* individualism.

The heroes of great culture are great leaders, and we have been privileged to work with many that we honor. They have made our job easy!

Conclusion:
Living Above the Line

The greatest glory in living lies not in never falling,
but in rising up every time we fall.

—Nelson Mandela

As we come to the conclusion of this mountain climb together, once again we draw on Mandela's wisdom. Sometimes we do fall off our "personal growth mountain." Sometimes this self-awareness peak seems too much, too steep, too hard. Yet another ascent looms before us and we have one of those "oh-not-this-*again*-I-thought-I-dealt-with-that-ages-ago" moments.

Like learning any new skill, growing into ATL thinking and behaviors can feel awkward at first. It takes time and focus. As we said in the introduction, rising up is about grace and kindness—for yourself just as much as for others. Give yourself time, space, and patience. We call it *going back to base camp*. After setting up multiple camps on a big mountain like Mount Everest, climbers retreat to base camp to rest and rejuvenate before the final summit bid.

Take a moment to find *your* base camp to rejuvenate and lighten the load. And if today felt like an uphill battle, stop, turn around, and see just how far you've climbed! Through time, as you become more and more able to see situations for either the potential triggers they might present for you or for the opportunity to practice specific ATL thinking and behaviors (and every day is full of those opportunities!), you will find yourself developing a deeper awareness of yourself and others that is more than worth the climb.

You're Installing a New Filter System in Your Heart and Mind

You're probably familiar with a phenomenon that happens when you buy something new—a car, sunglasses, computer, particular breed of dog or cat—and out of nowhere you start spotting that exact same thing over and over, everywhere, around your neighborhood or at work. It happens because your brain's filter system seeks out the familiar (that limbic system again!). Many people who have started the Heartstyles journey have told us they feel they have a new filter system in their mind and heart. They start to see instances of authentic and reliable behaviors, or controlling and avoiding ones, all around them (and within, too!) Day by day, month by month, and year by year, seeing life through the Heartstyles filter has helped them understand themselves and others better—to spot the behavior styles, both BTL and ATL, and notice opportunities to make new choices to transform their lives and those around them. The filter system goes to work and makes it easier for us.

Developing a refined filter takes time—and yet more patience—and requires us to put one foot in front of the other and trust the process (TTP!). Even though we will face setbacks, mistakes, disappointments, and the realities of life, we will find more of our best selves, our true north, the incredible people we can be by trusting the process, trusting ourselves, and trusting our potential.

Success is never final, failure is never fatal: it is the courage to continue that counts.

—WINSTON CHURCHILL

The True Summits of Your Life

In the world of mountain climbing, there's such a thing as a false summit. As you climb toward the summit, from a distance it appears to be the pinnacle of the mountain, your ultimate goal. But as you approach it, you finally see that it is in fact a smaller peak, not the true summit you were aiming for at all. Your hopes dashed, you may despair and even lack the will and desire to continue!

When your dreams, goals, desires, and aspirations are grounded in humility and love, they will bring you much joy, fulfillment, and satisfaction. If they are rooted in fear and pride, they will never be enough. They will entice you on, but give you nothing more than short-term fulfillment and leave you forever wanting more and more.

At those times when you ask yourself, *What on earth am I here for?* we encourage you to set your hearts on the true summits of life. We were created by love, for love. This is why we have hidden gifts, talents, and gold deposited in our hearts. We are here to make the world a better place. We are part of the solution, not part of the problem. When we seek wisdom, we are able to impact positively and leave a legacy.

Many an experienced mountaineer, for all their skill and knowledge, perishes and remains forever on the mountain. Some are overcome through pure accident, and others through lack of wisdom when ego-driven pride takes over.

From the summit of the mountain, train your heart to listen. Train your heart to accept. Train your heart to not be offended. Train your heart to seek wisdom to build your inner character strength.

When humility and love are in our hearts, wisdom will deliver answers. Wisdom's song is not heard in the spirit of intellectualism. Intellectualism stimulates the mind but does not move the heart. Instead, wisdom's lyrics can be heard in the heart of the humble. Their refrain is "it's not about you"! Wisdom will guide you to discover your authentic self and acquire the very things you're looking for. When wisdom wins your heart, and revelation breaks in, true pleasure enters your soul and you prosper. A revelation is to *reveal-action*. For the leader, wisdom will empower you with brilliant strategies to build, release, and love others—your organization, your family, and your community—to reach their destiny.

Four Seasons in Your Heart

How confusing life can be at times, full of what we call ups and downs. Most times wisdom allows us to look back at those ups and downs and see the good in them, what they taught us. And it seems to not matter how much you know, no matter how old you are, or how much wealth, success, and happiness you have, you still go through these *seasons of life*. Just as there is a cycle of four seasons in nature, there is also a cycle in life.

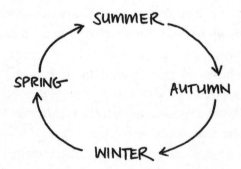

Summer is the bright time in life when things are going well. In that moment with the sun shining on us, it feels like it will last forever. But as we know, life will change: we change jobs, get married, have children, move state or country. We become more successful, and that creates more travel, which is disruptive to the family or relationships. We get a promotion. The company restructures. Something in life happens that we did not plan for. Sometimes the season is changing and we don't want to admit it—we just want to stay in summer and yet we're about to have a "fall."

Fall is when it can feel as though everything is falling into a black hole—when just a moment ago it was all going so well! And the truth is, life is always changing. It struck me when walking through Hyde Park in London one day that all the people who had stopped to take photos of the magnificent red and gold leaves were actually admiring something dying—and yet this death in nature produces amazing beauty. So it is with the changing seasons of life. Things die all the time. Attitudes change; opportunities, projects, and relationships come to an end. As humans we don't always see the beauty in it. We often see it as painful or frustrating. Yet if we try, we can see these seasonal changes as something not only beautiful but beneficial—something photo-worthy. Something worth noting in your journal.

Winter is the cocooning time, and it can feel dark and depressing. Even keen skiers tend to stay indoors more in the winter, and in our own "winter season" we go inside ourselves because we are going through change. The time between being a caterpillar and butterfly, the time of transformation, can be a dark place—but as we TTP we can remain in our winter season with more peace. Without faith in the process of change and development, we can become depressed, so be kind to yourself. Remind yourself that this is only a season and you will come out stronger for it. Remember, you have value—if you take two twenty-dollar notes to a bank, one looking crisp and new and the other crushed and worse for wear—they actually both have the *same value*. As do you when you're going through your winter season. Spring and summer are coming!

Spring is where we can see good things happening, though they might be only tiny things. We have been in caves where it is so dark you cannot see your hand in front of your face, yet the tiniest speck of light from a match or a flashlight has power over the darkness. So look for the tiny specks of hope, specks of light, specks of new growth. You might have a revelation in the shower and all of a sudden new buds and blossoms appear in your thought life. A thought in the shower, an idea in a dream, a friend's kind word, an offer or an opening comes our way to give us hope, when just last week there was nothing but winter. Amazing!

And then the next season comes. Summer again!

Faith and Moving *Beyond* the Line

He who loses money, loses much. He who loses a friend, loses much more. He who loses faith, loses all. May you all continue to have faith in someone and something.

—ELEANOR ROOSEVELT

Last, turning fear into faith will give you a strength beyond the self. The bottom line of below the line is a lack of faith. Here's a thought: without surrender to a higher power, we can reduce the world, the universe, and our minds to our own ego-self, and that can be a small place!

We are not human beings having a spiritual experience; we are spiritual beings having a human experience.

—POPULARIZED BY DR. WAYNE DYER AND ATTRIBUTED TO PIERRE
TEILHARD DE CHARDIN, FRENCH PHILOSOPHER AND JESUIT PRIEST
(1881–1955)

There are times our hearts have the opportunity to yield *beyond* the line, to faith—faith in humility and love, faith in the product or service, faith in people, faith in the team, faith in supportive family and friends. We also hope you seek and find your spiritual faith, the kind of faith in a higher power that you surrender you heart to. This ultimate place of surrender is the ultimate place of peace that passes knowledge.

Like you, the two of us have done life and are doing life still. We're in this journey of life together, so we have made our mistakes and hopefully learned from them. Both of us are fortunate to have found our faith, to live by that faith, and surrender to it daily. Most of all, when our human BTL fear and pride come to deceive us and take us captive, we can rise up above the line and even beyond the line (as we are human, too, sometimes it takes a bit of teeth gnashing to start!).

Remember, we're only as good as our next ATL moment, so let's run toward it with our arms open wide. Embrace who we are and our journey of *becoming*!

Message from a Friend

In closing, a dear friend of ours, who spent hours and hours of his life helping us with the very early development of Heartstyles, passed away from pancreatic cancer at fifty-two. As we were finishing this book, his passing was exactly ten years ago. Before our friend passed, we asked him to write a list of lessons for us to treasure. To honor him and leave you with wisdom from a person whose legacy lives on through this book, we share with you his list below. Thank you, Phil.

1. The breath we take in on a moment-by-moment basis is the free breath of life.

2. Try to have good days and bad moments. Life is too short to have a bad day.

3. Keep your disagreements short because you will never know how long you have to rectify them.

4. Listen to those you care for, because their journey is just as important as yours.

5. Share with those you care for, because they need to hear what is in your heart.

6. Live up to the words of encouragement spoken over you by others.

7. Focus on the critical few things in any situation and not the insignificant many.

8. Remember to thank those who care for you.

9. Spend time with each other smelling the roses, watching a sunrise or sunset. Although they appear every day, they are never the same.

10. Pray over your wife, husband, family member, or friends in the way you would like them to pray over you.

11. Aim for the heart in any situation, because people don't care how much you know until they know how much you care.

12. Try walking in each other's shoes for a day and when you return to your own, they will feel much more comfortable.

13. When trying to resolve a problem, approach the solution from a different level of thinking.

14. Healing doesn't mean you forget; it means you don't hurt anymore.

15. Don't go into battle in someone else's armor—live your own experience.

16. It is easy to go down the "wrong thinking" track, but remember, no matter how far down that track you go, God will always walk with you.

17. Help me transform my future so I don't repeat my past.

18.   The reason we are losing the battle is because we think we are. Stop and think what it would be like to win the battle.

19.   *Thank you* never seems enough, but when it comes from deep in the heart, words cannot explain that it is.

20.   Don't think about it, do it—because by the time you have put your thoughts into action, unrecoverable time has passed you by.

So, wonderful people, go and live and lead with heart, above the line and beyond the line, making a difference to the world around us. One heart at a time. Peace.

Acknowledgments

Tommy Spaulding—For your heart in wanting to see us put our work into a book and introducing us to your agent, Michael Palgon. Thank you for the foreword and, most of all, leading the way with your book *The Heart-Led Leader.*

Michael Palgon—Our (intrepid!) agent. Thank you for your authentic coaching with compassion as you guided these two Aussie beginners on this mountain climb of book writing and publishing.

Lari Bishop and Sally Collings—Our ghost writers. There is no possible way we could have done this without you. Thank you for your patience with us, having to navigate this big concept topic and two authors! Thanks for believing in what we are doing as you have professionally engaged with us.

Hollis Heimbouch—For believing in this message, seeing how it can help people, and taking on first-time authors with Harper-Collins. Aussies who live in London are not necessarily as easy to take on as someone living in America, and you still believed our message is global. Thank you!

Stephanie Hitchcock—Thank you for passionately getting involved in this material, editing and making format suggestions, and being so great to work with in making this dream become a reality.

Our clients—Over the past thirty years there are so many clients to thank, past and current. However, when pioneering a business with new concepts (we call them start-ups these days), one needs early adopters. There are three organizations we would like to honor with all our heart. Our first client in 1987, Sykes Equipment Finance, since 1989 Kennards Self Storage, and since 1992 Yum! Brands (KFC, Pizza Hut, Taco Bell). They started with our original brand Achievement Concepts and are still working with us and developing people and culture in their organizations. A *very* special and heartfelt thanks to Yum! Brands for believing in developing Heart in Business, for being such early adopters of Heartstyles globally, leading to the translation of our work into twenty-five languages. This is a leading with heart organization.

To our client CEOs and senior leaders—Thank you for who you are, for your heart to want to build great cultures and people, and for being brave enough to go on the Heartstyles journey yourselves, leading by example through your courage and vulnerability with your people. It is true heart-led leadership. Thank you for trusting us when we challenged you to move *way out* of your comfort zones—look at what you've achieved since and the legacy you are leaving! A special mention to Jens Hofma, CEO of Pizza Hut Restaurants UK, for being such an early adopter of this philosophy incorporated into the corporate sector, and for your passionate leadership and belief that everyone deserves a chance to develop.

Participants—The countless number of people who have trusted us with your hearts and lives, and allowed us to share this wisdom with you and in turn learn from you. All those of you who so courageously and enthusiastically answered our call for stories for this book. You have so graciously told us your stories and allowed us to weave them into the book, so that others can be helped in their own journey of transformation. Thank you.

Our team—So many over thirty years, but as we publish this book, we want to acknowledge our gratitude to our current senior team who have taken over the day-to-day operations of the Heartstyles business to allow us time to write, and to our wider team—past and present—who have joined the cause and served with such passion. Our "next generation" team have become extraordinary speakers and facilitators in their own right, sharing this message with commitment and passion. We know you are taking this beyond what we could do, and our vision and brand is safe in your hands. Andy Read, who has spent years taking this message to village chiefs and leaders throughout Africa, truly showing this message transcends across *all* people and cultures.

Stephen would like to say: As a leader who has been BTL at times, *having to learn that which I have been given to teach*, I am deeply grateful for the grace, love, forgiveness, faith, prayer, and honor my team have extended to me. My deepest gratitude to you.

Our accredited associates and certified practitioners—Thank you for your passion in carrying the heart of Heartstyles personally and in your facilitation. Through you we can reach the millions of people for whom this book is written.

Family—Mara's late parents, Laura (Moo) and Lorenzo (Papa), and Stephen's late parents, Oscar (Klem) and Joyce. To Nathan and Tamara—we love you with all our heart, thank you for loving us through life and for teaching us so many lessons of living above the line. To our extended family, we hope life continues to go well for you. Our "adopted spiritual" sons and daughters, we love you. Family and friends who are in the afterlife, you remain in our hearts, and your absence is further encouragement for us to make a difference on this planet while we can, and to finish well.

Friends—Just too many to list, you know who you are. Thank you for walking with us through so many ups and downs, *'cause that's what friends are for* . . .

Other authors, theorists, philosophers, and pastors that we have learned from. Nothing new under the heavens!

God, our heavenly Father—The most secure fact we have in our lives is that God loves us—as God is love, unconditional love. In our fearful times we have been able to posture our hearts to this love for our deepest security. Let love live.

Stephen says: My darling wife, who has climbed mountains with me, literally, all over the world. We have risked our lives together hanging on a bit of rope. However, the risk you have taken in climbing this mountain, the Heartstyles mountain, is greater and has taken such faith when year after year, over two decades, you have breathed life into a dream. You have invested your time, energy, intellect, amazing talent, and life savings into this dream, not knowing for many years if it would ever come to anything. You have guided the translations of twenty-five languages, given countless hours, and now we have reached a summit with our vision. From the summit we see the world together. You are my angel from heaven. I love you.

Mara says: My darling, you changed my life. You brought me a family and have been my "irritant" that has supported, sometimes pulled, and a lot of time pushed (in an above the line way!) me into my own development! Thank you for the way you love me, for being my rock, my love, and my partner in this crazy life. There is no greater privilege than seeing people step into their courage to set themselves free and become a better version of themselves. Thank you for giving me that opportunity. *Ti amo per sempre.*

You, the reader—Thank you for coming on the journey to learn and challenge yourself to become the best version of you. If we can all just do a little bit of that, we can change the world—one heart at a time.

Appendix

The QR code below is your link to access workbook pages, view video learning, and download our app to connect with us on your mobile device. When you see an asterisk * in the main body text of this book, that means you can access additional material through this QR code.

www.heartstyles.com/book

Notes

CHAPTER 1: THE FOUR UNIVERSAL PRINCIPLES THAT SHAPE YOUR LIFE

37 "showing fear is considered weakness": Kate Murphy, "Outsmarting Our Primitive Responses to Fear," *New York Times*, October 26, 2017.

38 "companionate love": "What's Love Got to Do with It? A Longitudinal Study of the Culture of Companionate Love and Employee and Client Outcomes in a Long-term Care Setting" (*Administrative Science Quarterly*, 2014, vol. 59, issue 4).

39 a definition of authentic humility: Rich Barlow, "Studying the Benefits of Humility" (*BU Today*, March 27, 2017).

CHAPTER 2: TRIGGERS, TEMPLATES, AND TRUTHS

47 "Human freedom": Rollo May, *The Courage to Create* (New York: W. W. Norton, 1975).

CHAPTER 3: VOIDS, WOUNDS, AND THE GOLD WITHIN THE HEART

67 the consequences of early relationships for children's emotional development: Jallu Lindblom et al., "Early Family Relationships Predict Children's Emotion Regulation and Defense Mechanisms" (SAGE Open, December 6, 2016).

71 "You may not control all the events": Maya Angelou, *Letter to My Daughter* (New York: Random House, 2008).

CHAPTER 6: TRADE UP! HOW TO RESIST BEING PULLED BELOW THE LINE

146 amygdala hijack: Daniel Goleman, *Emotional Intelligence: Why It Can Matter More Than IQ* (New York: Bantam Books, 1996).

156 new habit: Phillippa Lally et al., "How Are Habits Formed: Modelling Habit Formation in the Real World" (*European Journal of Social Psychology*, October 2010, vol. 40, issue 6, 998–1009).

CHAPTER 7: KNOW WHO I AM—AUTHENTIC AND TRANSFORMING
162   Even a *Forbes* article: Laura Entis, "This is the No. 1 Thing These CEOs Look For in Job Candidates," *Forbes*, March 26, 2017 https://fortune.com /2017/03/26/ceos-ideal-job-candidates/.

176   "raw truth and openness": Brené Brown, *Daring Greatly: How the Courage to Be Vulnerable Transforms the Way We Live, Love, Parent, and Lead* (New York: Avery, 2015).

176   "beautiful mess effect": Anna Bruk et al., "Beautiful Mess Effect: Self-Other Differences in Evaluation of Showing Vulnerability" (*Journal of Personality and Social Psychology*, 2018, vol. 115, issue 2, 192–205).

CHAPTER 8: KNOW WHERE I'M GOING—RELIABLE AND ACHIEVING
188   "captivated with purpose": Bob Goff, *Love Does: Discover a Secretly Incredible Life in an Ordinary World* (Nashville, TN: Thomas Nelson, 2012).

CHAPTER 9: CONNECTING WITH OTHERS—RELATING AND ENCOURAGING
200   emotional bank account: Stephen Covey, *The 7 Habits of Highly Effective People* (New York: Free Press, 1989).

211   "Technoference": Brandon T. McDaniel and Jenny S. Radesky, "Technoference: Parent Distraction with Technology and Associations with Child Behavior Problems" (*Child Development*, 2018, vol. 89, issue 1, 100–109).

CHAPTER 10: GROWING WITH OTHERS—DEVELOPING AND COMPASSIONATE
218   growth mindset: Carol S. Dweck and David S. Yeager, "Mindsets That Promote Resilience: When Students Believe That Personal Characteristics Can Be Developed" (*Educational Psychologist*, 2012, vol. 47, issue 4, 302–14).

220   "emotional bank account": Stephen Covey, *The 7 Habits of Highly Effective People.*

224   separating the *person* from the *behavior*: Kenneth Blanchard and Spencer Johnson, *The One Minute Manager* (San Francisco: Jossey-Bass, 1996).

230   Peter Principle: Dr. Laurence J. Peter and Raymond Hull, *The Peter Principle* (New York: William Morrow & Co, 1969).

CHAPTER 11: BUILDING ABOVE THE LINE WORK CULTURE
243   millennials want community: *How Millennials Want to Work and Live* (Gallup Report, 2016).

254   "what got us here, won't get us there": Marshall Goldsmith, *What Got You Here, Won't Get You There* (London: Profile Books, 2008).

258   "At the heart of vulnerability": Patrick Lencioni, *The Advantage* (San Francisco: Jossey-Bass, 2012).

Index

About the Authors

STEPHEN KLEMICH is a longtime leadership consultant and speaker, and the CEO and founder of Heartstyles. His multinational clients have included KFC, Pizza Hut, Taco Bell, Unilever, American Express, and PricewaterhouseCoopers.

MARA KLEMICH, PhD, is a consulting psychologist and the co-founder of Heartstyles.

They are Aussies living in London.